GLOBALISATION OF WATER
OPPORTUNITIES AND THREATS OF VIRTUAL WATER TRADE

T0362237

Globalisation of water
Opportunities and threats of virtual water trade

DISSERTATION

Submitted in fulfilment of the requirements of
the Board for the Doctorate of Delft University of Technology
and the Academic Board of the UNESCO-IHE Institute for Water Education
for the Degree of DOCTOR
to be defended in public
on 24, April 2006 at 15:30 hours
in Delft, The Netherlands

by

Ashok Kumar CHAPAGAIN

born in Dingla, Bhojpur, Nepal
Master of Science, UNESCO-IHE, The Netherlands

CRC Press
Taylor & Francis Group
Boca Raton London New York

CRC Press is an imprint of the
Taylor & Francis Group, an **informa** business
A TAYLOR & FRANCIS BOOK

This dissertation has been approved by the Promotors
Prof. dr. ir. H. H. G. Savenije, UNESCO-IHE / TU Delft, The Netherlands
Prof. dr. ir. A. Y. Hoekstra, University of Twente, The Netherlands

Members of the Awarding Committee:

Chairman	Rector Magnificus TU Delft, The Netherlands
Vice-chairman	Rector, UNESCO-IHE, The Netherlands
Prof. dr. ir. H. H. G. Savenije	TU Delft/UNESCO-IHE, The Netherlands, promotor
Prof. dr. ir. A. Y. Hoekstra	University of Twente, The Netherlands, promotor
Prof. dr. J. A. Allan	SOAS, University of London, UK
Prof. dr. H. J. Gijzen	UNESCO-IHE, The Netherlands
Prof. dr. ir. P. van der Zaag	TU Delft/UNESCO-IHE, The Netherlands
Prof. dr. ir. A. van der Veen	University of Twente, The Netherlands

CRC Press
Taylor & Francis Group
6000 Broken Sound Parkway NW, Suite 300
Boca Raton, FL 33487-2742

© Ashok K. Chapagain, 2006

CRC Press is an imprint of Taylor & Francis Group, an Informa business

No claim to original U.S. Government works

ISBN-13: 978-0-415-40916-2 (pbk)

Visit the Taylor & Francis Web site at
http://www.taylorandfrancis.com

and the CRC Press Web site at
http://www.crcpress.com

To my late father who could not witness the final outcome of this research.

Abstract

Where the river basin is generally seen as the appropriate unit for analyzing freshwater availability and use, it becomes increasingly important to put freshwater issues in a global context. The reason is that international trade of products brings along international and intercontinental transfer of large volumes of water in virtual form. Various water-scarce countries intentionally import 'virtual water' in order to reduce the pressure on own domestic water resources. 'Virtual water' is understood here as the volume of water that is used to produce a commodity. The objective of the research is to analyse the opportunities and threats of international virtual water trade in the context of solving national and regional problems of water shortages. Central questions addressed in the study are: What are the fluxes of virtual water related to the international trade of products? Is the import of virtual water a solution to water-scarce nations or merely a threat of becoming water dependent? Can the international trade of products be a tool to enhance water use efficiency globally, or, is it a way of shifting the environmental burdens to a distant location? To understand the global component of fresh water demand and supply, a set of indicators has been developed. The framework thus developed has been applied to different case studies.

Virtual water flows between nations have been estimated from statistics on international product trade and the virtual water content per product in the exporting country. The calculated global volume of virtual water flows related to the international trade in commodities is 1625 Gm^3/yr. About 80% of these virtual water flows relate to the trade in agricultural products, while the remainder is related to industrial product trade. An estimated 16% of the global water use is not for producing domestically consumed products but products for export. With increasing globalisation of trade, global water interdependencies and overseas externalities are likely to increase. At the same time liberalisation of trade creates opportunities to increase global water use efficiency and physical water savings.

Many nations save domestic water resources by importing water-intensive products and exporting commodities that are less water intensive. National water saving through the import of a product can imply saving water at a global level if the flow is from sites with high to sites with low water productivity. The research analyses the consequences of international virtual water flows on the global and national water budgets. The assessment shows that the total amount of water that would have been required in the importing countries if all imported agricultural products would have been produced domestically is 1605 Gm^3/yr. These products are however being produced with only 1253 Gm^3/yr in the exporting countries, saving global water resources by 352 Gm^3/yr. This saving is 28 per cent of the international virtual water flows related to the trade of agricultural products and 6 per cent of the global water use in agriculture. National policy makers are however not interested in global water savings but in the status of national water resources. Egypt imports wheat and in doing so saves 3.6 Gm^3/yr of its national water resources. Water use for producing export commodities can be beneficial, as in Cote d'Ivoire, Ghana and Brazil, where the use of green water resources (mainly through rain-fed agriculture) for the production of stimulant crops for export has a positive economic impact on the national economy. However, the use of 28 Gm^3/yr in Thailand for the production of rice for export is at the cost of additional pressure on its blue water resources. Importing a product which has a relatively high ratio of green to blue virtual water content saves global blue water resources that generally have a higher opportunity cost than green water.

The use of virtual water transfers as an alternative to large scale inter-basin real water transfers has been analysed in a case study for China. North China faces severe water scarcity – more than 40% of the annual renewable water resources are abstracted for human use. Nevertheless nearly 10% of the water used in agriculture is applied for producing food exported to South China. To compensate for this 'virtual water flow' from North to South and to reduce water scarcity in the North, the huge South-North Water Transfer Project is currently being implemented. This paradox, transfer of huge volumes of water from the water-rich south to the water-poor North versus transfer of substantial volumes of food from the food-sufficient North to the food-deficit South, is receiving increased attention, but the research in this field stagnates at the stage of rough estimation and qualitative description. The current study quantifies the volumes of virtual water flows between the regions in China and places them in the context of water availability per region. North China annually exports about 52 billion m^3 of water in virtual form to South China, which is not much more than the maximum proposed water transfer volume along the three routes of the Water Transfer Project from South to North.

In order to quantify and visualise the effect of consumption in a nation on the globe's water resources, the study uses the water footprint concept. The water footprint of a nation is the total volume of fresh water that is used to produce the goods and services consumed by the nation. A water footprint is expressed in terms of the volume of water use per year. The internal water footprint is the water use in the country considered while the external water footprint represents the water use in other countries. The study has quantified the water footprint for each nation of the world for the period 1997-2001. The USA has an average water footprint of 2480 m^3/cap/yr, while China has an average footprint of 700 m^3/cap/yr. The global average water footprint is 1240 m^3/cap/yr. There are four major direct factors which determine the water footprint of a country: volume of consumption (related to the gross national income); consumption pattern (e.g. high versus low meat consumption); climate (growth conditions); and agricultural practice (water use efficiency).

The global water footprint of coffee and tea consumption has been elaborated with an example for the Netherlands with the underlying aim of contributing to figures that can be used for raising awareness on the effects of consumption patterns on the use of natural resources. The standard cup of coffee and tea in the Netherlands costs about 140 litres and 34 litres of water respectively, by far the largest part for growing the plant. The large volume of water to grow the coffee plant comes from rainwater, a source with less competition between alternative uses than in the case of surface water. For the overall water need in coffee production, it makes hardly any difference whether the dry or wet production process is applied, because the water used in the wet production process is a very small fraction (0.34%) of the water used to grow the coffee plant. However, the impact of this relatively small amount of water is often significant. First, it is blue water (abstracted from surface and ground water), which is sometimes scarcely available. Second, the wastewater generated in the wet production process is often heavily polluted.

Similar to coffee and tea, the consumption of a cotton product is connected to a chain of impacts on the water resources in the countries where cotton is grown and processed. The study has estimated the 'water footprint' of worldwide cotton consumption, identifying both the location and the character of the impacts. The research distinguishes between three types of impact: evaporation of infiltrated rainwater for cotton growth (green water use), withdrawal of ground- or surface water for irrigation or processing (blue water use) and water pollution during growth or processing. The latter impact is quantified in terms of the dilution volume necessary to assimilate the pollution. For the period 1997-2001 the research shows that the worldwide consumption of cotton products requires 256 Gm^3 of water per year, out of which about 42% is blue water, 39% green water and 19% dilution water. Impacts are typically cross-border. About 84% of the water footprint of cotton consumption in the EU25 region is located outside Europe, with major impacts particularly in India and Uzbekistan. Given the general lack of proper water pricing mechanisms or other ways of transmitting production-information, cotton consumers have little incentive to take responsibility for the impacts on remote water systems.

The research shows that international trade has indirectly enhanced the global water use efficiency and helped to address the national water scarcity in some water-poor countries by saving national water resources. However, this was possible at the cost of increased water dependencies between nations. The existing indicators of water use are not sufficient to address the effect of consumption on water resources. It is proposed to use the concept of water footprint to understand the real appropriation of water by a nation and also to understand the chain of impacts on global water resources as a result of local consumption. The future trade negotiations should undertake the notion that trade is not only a tool of global economic development; it can also be a means of externalising the water footprint and thus shifting environmental burdens to distant locations.

Contents

Chapter 1

Introduction

Water shares a number of characteristics with other natural resources. Yet, not all resources share water's significance as the basis for all forms of life. Though water is abundant globally, albeit unevenly distributed in time and space, it is becoming scarcer in two senses. First, the global population is increasing rapidly, mostly in developing regions of the world, and second, the quality of the available fresh water is deteriorating with increased pollution from human activities. As pointed out by Cosgrove and Rijsberman (2000), the crisis will become more serious under a business as usual scenario and problems of water shortages and pollution will intensify unless effective and concerted actions are taken.

Water management contributes directly and indirectly, in a number of ways, in achieving the Millennium Development Goals established by the UN General Assembly Millennium meeting in 2000 (UNESCO-WWAP, 2003). WHO/UNICEF (2000) has estimated that today 1.1 billion do not have access to clean drinking water and 2.4 billion have no provision for sanitation. It has been estimated that nearly 7 billion people in sixty countries will live in water-scarcity by 2050 (UNESCO-WWAP, 2003). Even under the lowest projection, still nearly 2 billion people in forty-eight countries will struggle against water scarcity in 2050 and it is predicted that at least one in every four people in 2050 is likely to be in countries which will be affected by chronic shortages of fresh water (UNESCO-WWAP, 2003).

Many rivers, lakes and groundwater resources are becoming increasingly polluted from domestic and industrial waste disposal, return flows from agricultural fields where use of chemical fertilisers and pesticides are common practice. Water quality is also deteriorating as a result of the sediments from human induced erosion, increased salinity of ground water bodies as a result of saltwater intrusion, oil-spillages from river traffic etc.

The three major factors causing increasing water demand over the past century are population growth, industrial development and the expansion of irrigated agriculture. Falkenmark and Rockström (2004) observe that if we are worried about water scarcity problems today one can imagine how big will it be in the next 50 years as we will need an additional amount of water that is almost three times the amount presently used in irrigated agriculture to produce food for the additional 3 billion people in the next 50 years.

The common practice to meet the demand in time and place is by storing during peak flow periods and making it available in low flow periods. These engineering solutions are a form of supply management. Emphasis on water supply, coupled with weak enforcement of regulations, has limited the effectiveness of water resource management, particularly in developing regions (UNEP, 2002). With the examples of Saudi Arabia, Libya and USA's Ogallalla aquifer, Postel (1992) argues that the economic fate of these regions is unsustainable, being linked to a non-renewable water supply as these regions base their economic development on pumping of non-renewable fossil groundwater. These days the importance of demand management has become more important than ever.

The potential of demand management is most evident in areas where water use is wasteful in both urban water use and irrigated agriculture. However, in areas where water truly falls short and where the rains are the common and main source of supply, it may be erroneous to believe that demand management may contribute more than locally in addressing the water problems, especially as the water consumption for food production is climatically driven (Falkenmark and Lundqvist, 1997). Falkenmark and Lundquist (1997) further argue that irrespective of the management, the minimum water requirements are basically determined by the evaporative demand of the atmosphere and the length of the growing season for the crops selected.

There is a direct correlation between population growth and the increase in freshwater consumption. As human populations grow, as standards of living improve, as industrial productions expand, and as the need for food in dry areas increases the need for irrigation, water supply systems are increasingly likely to become both objectives of military action and instruments of war (Gleick, 1991). Gleick (1991) argues that the characteristics that make water likely to be a source of strategic rivalry both nationally and internationally include the degree of scarcity and the degree to which the water in a river basin is to be shared between different sectors, between urban and rural areas, and between different basin units (municipalities, states, nations).

Globalisation of trade

With the explosion of information and communication systems, the dismantling of trade barriers and the increasing economic power of trans-national corporations the international trade is increasing rapidly over the past quarter of a century. With the voyage of Columbus, the door was opened for the 450 years of European colonialism, and it was this centuries-long imperial era that laid the groundwork for today's global economy (Ellwood, 2001). Global trade expanded rapidly during the colonial period as European powers imported raw materials from their dominions: furs, timbers and fish from Canada; slaves and gold from Africa; sugar rum and fruits from the Caribbean; coffee, sugar, meat, gold and silver from Latin America; opium, tea and spices from Asia. European power stacked wealth from their overseas colonies but part of it also went back as an investment – into railways, roads, ports, dams and cities (Ellwood, 2001). There are different distinct landmarks in the history of globalisation as pointed out by Rennen and Martens (2003) depending on the chosen perspective, such as the political perspective (discovery of America in 1492), the economic perspective (foundation of the Dutch United East Indies Company in 1602), the technological perspective (invention of steam engine in 1765), or the environmental perspective (club of Rome's limits to growth in 1972).

Globalisation is a highly contested concept and definitions of the concept abound. It is a complex process spanning from economics to the cultural, environmental, and political domain. The views on globalisation are many times biased and the images are often painted to suit the philosophy followed. Even the protests are not to deny the phenomenon, but mostly focused on the way it should be defined and directed (Rennen and Martens, 2003). With the term globalisation, a majority of people think about the expanding international trade in goods and services based on the concept of comparative advantage (Ellwood, 2001). Globalisation today is different from the historic globalisation. Since the early 1970s the change became evident with the collapse of rules to manage global trade set up at the end of World War II. With the emergence and global influence of free-market governments in the UK and the US, companies became free to move their operations anywhere in the world to minimise costs and maximise returns to investors. Besides having an initial boost up in the national economies, the collapse of the East Asian currencies in July 1997 shows the risks included in the process (Ellwood, 2001).

International interaction, rather than isolation, has been the basis of economic progress in most places of the world. Watkins and Fowler (2002) believe that the great rewards of globalised trade have come to some but not to others. The strong opponents of globalisation think that the current regime advances a primitive winner-takes-all competition that inexorably widens the gap between rich and poor. They strongly believe that large-scale industrialised agriculture and unfair global trade is the root cause of poverty and environmental degradation (Cavanagh and Mander, 2002). Agricultural subsidies, export dumping and trade barriers are some key issues in most of WTO meetings (Oxfam, 2003; Watkins and Fowler, 2002).

Globalisation may bring uncertainties in markets but it also opens doors for new opportunities (Daveri et al., 2003). Concentrating on just one side of the coin gives a misleading picture of globalisation. As pointed out by Dicken (1992) the challenge should be to meet the material needs of the world community as a whole in ways which reduce, rather than increase, inequality and which do so without destroying the environment.

Efforts have been made to address complex societal and environmental aspects of globalisation, but research is still in its infancy when it comes to water issues. It is common to find intense discussions on the pros and cons of globalisation related to the international trade of products. The discussions are generally about the use and benefit of land, capital and human resources. Hardly any water issue is at the forefront.

Globalisation of water

Although no global government does exist, a number of global and regional organisations established to address trans-boundary water issues constitute an emergent system of global governance reflecting increased political coordination among governments, intergovernmental organisations and trans-national social movements. In this process a common purpose and goals via agreed rules, values and principles are worked towards. There are presently 261 international river basins, and 145 nations have (part of their) territory in shared basins covering about 50% of the land surface of the globe and including more than 40% of its population (UNESCO-WWAP, 2003). International water regimes came into being with the establishment of the Mekong River Commission in 1957. The establishment of various international commissions to deal with water issues is one more indication of the globalisation of water issues. For example, Angola, Botswana and Namibia established the Permanent Okavango River Basin Water Commission in 1994 to co-ordinate and collaborate on the sharing of the basin's water resources. The establishment of the Nile Basin Initiative in May 1999 with a vision to achieve sustainable socio-economic development through the equitable utilisation of, and benefit from the common Nile Basin water resources is another example of cross-border partnership for managing a river basin. However, all these efforts are only a step further down the road of managing the water resources extending the administrative borders from the national to the basin scale. They still seek the solution within the basin boundary.

There are a number of schemes to transfer water from surplus to deficit regions, occurring mostly within national or political borders. Now proposals for bulk water transfers are being made at international, and even global, levels (Gleick *et al.*, 2002a). Some of these large scale transfer projects either in conception or being implemented are the Inter Basin Water Transfer Link Project of India, the South-North Water Transfer Project of China and the Lesotho Highlands Water Project of Lesotho and South Africa. Large-scale bulk trading of fresh water has now become an issue in international trade negotiations and disputes. Some of these transfers (China and India) are solely being implemented at national scale, though the project in India has implications on the shared basin of the Ganges.

The trade in bottled water was 57000 million litres in 1996 and is projected to be around 143,800 million litres per year in 2006 (Gleick *et al.*, 2002a). In recent years, several efforts have been made to implement standard rules governing international trade such as the General Agreement on Tariffs and Trade (GATT). GATT provides the basic legal architecture that governs international trade for the member countries of the World Trade Organization (WTO). Gleick *et al.* (2002a) conclude that as great uncertainty continues to revolve around the legal interpretation of international trade agreements in the context of globalising water resources and as it has economic cost to transfer water in bulk, large-scale, long-term bulk exports of water across international borders are unlikely.

Another trend in the global water arena is the process of privatisation. Treating water as an economic good, and privatising is not a new idea. However, the extent of the privatisation effort and the awareness to the pros and cons of these processes is higher than ever. The potential advantage of privatisation is greatest where governments have been weak and have failed to meet basic water needs. The risk of privatisation however is also largest if governments are weak and unable to provide the oversight and management functions necessary to protect public interests. In many parts of the world, the rights to freshwater in rivers, streams, and lakes are being sold to giant transnational corporations which will ultimately charge users for every drop of water for their short term profits (Cavanagh and Mander, 2002). Gleick *et al.* (2002b) argue that any effort to privatise or commodify water should be accompanied by formal guarantees to respect certain principles and support specific social objectives such as: 'continue to manage water as a social good', 'use sound economics in water management', and 'maintain strong government regulation and oversight'.

Since water has been defined as a tradable commodity by both the North American Free Trade Agreement (NAFTA) and the World Trade Organisation (WTO), and as water services are also labelled as a commodity in the new General Agreement on Trade in Services (GATS), under a new category called 'Environmental Services', Cavanagh and Mander (2002) believe that once water is privatised, commodified, and put on the open market, it is not available to everyone who needs it but only to those who pay. As such, privatisation of water is becoming a subject of protest around the globe.

Efficient use of water resources

To address present and possible future problems of water shortages, the optimal use of the global water resources is becoming a focal point of discussion in the water arena. Efficiency can be measured either in economic sense or in physical sense. A resource might be used for more beneficial use than the existing one which can be measured with economic tools. Economic efficiency means that marginal cost of water use in a particular case does not exceed the marginal benefit. This means that the allocation of water to different uses is taken based on the economic returns and that the volume of use in a particular type of use is such that marginal benefit equals the marginal cost.

Hoekstra and Hung (2002; 2005) distinguished three different levels of water use where decisions can be made to enhance the water use efficiency, namely local level, basin level and global level. One can use a different crop or minimise wasteful use of water in agriculture or use more water efficient technology to achieve the same output, thus, enhance the local water use efficiency. Efficiency can be enhanced at basin level by allocating water to uses with higher marginal benefit. At global scale the efficiency can be enhanced by producing at more favourable sites. At all three levels one can look at how to increase physical efficiency, that means use less water to produce the same output (less cubic meter of water per kg or dollar of production) or at how to increase economic efficiency, that means optimise total output given the full set of resources including water. Sustainable management of water resources requires systemic, integrated decision-making that recognizes the interdependence of decision-making at all three levels (Gallopín and Rijsberman, 2000).

In this thesis, the global water use efficiency is interpreted in physical sense, so that if we speak of increasing global water use efficiency we refer to the fact that globally the same output is produced with lower water input. Global water use efficiency can be enhanced by importing from regions where the water productivity (i.e. output per cubic meter of water) is higher to regions with low water productivity.

The solution to water scarcity problems can be tackled at two distinct stages (Figure 1.1). Either one can produce the same using less water, thus using resources more efficiently, or make choices of consumption such that total demand itself is lower. Where supply management focuses on supplying additional water to fulfil additional demand, water demand management focuses on reducing water demand (in both production and consumption stage) to produce the same or even less if demand was beyond the level of the economic optimum. Economists generally see water pricing as the main (sole) instrument of demand management and this is a major pitfall (Savenije and van der Zaag, 2002). Demand management has different instruments such as: quota, water use licences, tradable water rights, user charges, subsidies, grants and penalties. One form of demand management can also be to import a water intensive good instead of producing locally. Importing products instead of producing where it is consumed means not only the import of products but also the import of 'virtual water'.

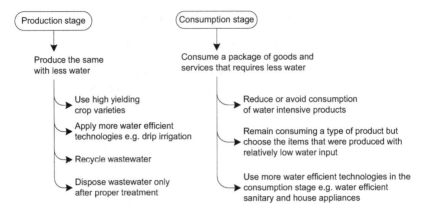

Figure 1.1. Solutions to water scarcity problems can be identified at two distinct stages.

Virtual water trade

The term 'virtual water' has been introduced by Tony Allan around 1993 (Allan, 1993; 1994). It is defined as the volume of water required to produce a commodity or service (Allan, 1998b; 1999b; Hoekstra, 1998). When there is a transfer of products or services from one place to another, there is little direct physical transfer of water (except the water content inside the product which is quite insignificant in terms of quantity). There is however a significant transfer of virtual water. From a country's perspective, Haddadin (2003) has defined this water also as 'exogenous water'. It is sometimes also known as ultra-violet water while different fluxes of water are given colours such as blue, deep blue, green, white, grey (Savenije, 2004).

More precisely the term virtual water can be defined with two distinct approaches. One is from the production point of view and another is from the use point of view (Hoekstra, 2003). The first approach quantifies virtual water as the real water used for the production of the commodity. It is production site specific as it depends on the production conditions, including place and time of production and water use efficiency. In the second approach, the virtual water content is defined as the amount of water that would have been required to produce the product at the place where the product is used. Hence it is use site specific. The first definition is useful if we are interested in how much water was *really* used to make a product, for instance for estimating the impact of the product on the environment. The second definition is useful if we think how much water a country can save by importing a commodity instead of producing it domestically.

In the second approach to the definition of 'virtual water' a difficultly arises if a product is imported to a place where the product cannot be produced, for instance due to climate conditions. What for instance is the virtual water content of rice in the Netherlands, where rice is not being produced (due to climatic conditions) but imported only? In this case, Renault (2003) proposes to look at the virtual water content of a proper substitute of the product considered. If the definition of virtual water content is approached in this way, one can even start arguing that seawater fish contains virtual (fresh) water even though this fish doesn't depend on fresh water at all. In the same line, Renault (2003) proposes the principle of nutritional equivalence, which provides a means to compare food products based on their nutritional values.

As there is a general trend of increase of water productivity in time, the virtual water content of commodities is time dependent (Renault, 2003). Thus, it is quite important whether one is looking at past or future virtual water. As a consequence, virtual water is neither constant in space nor in time.

Allan (1998b) explains why there has been no war over water in the Middle East, even though many economies in arid regions have only half the water they need. He states that the economic system has solved the water supply problem of the regions via virtual water trade. If one country exports a water-intensive product to another country, it exports water in virtual form. In this way some countries support other countries in their water needs. For water-scarce countries it could be attractive to achieve water security by importing water-intensive products instead of producing all water-demanding products domestically. Reversibly, water-rich countries could profit from their abundance of water resources by producing water-intensive products for export. Trade of real water between water-rich and water-poor regions is generally impossible due to the large distances and associated costs, but trade in water-intensive products (virtual water trade) is realistic. For a large country or a country with different climatic zones, the concept is equally applicable to improve the regional efficiencies within the country itself.

Allan (1998b) explains how the problem of water scarcity in a watershed can be very well addressed by taking an international economic perspective. The global economic system appears to be very important with respect to bridging local periodic droughts (Allan, 1999a; 1999b). Allan explained why despite the Middle East needing by 1990 twice as much water as available to meet its overall strategic needs, and despite the demographically driven trend which will mean that the region will need four times as much water as currently available by the third decade of the twenty-first century, it has become clear that the region can balance its water budget by importing 'virtual water'. At present, the MENA (Middle East and North Africa) region imports each year a volume of water equivalent to the annual flow of Nile into the region (Allan, 2001b).

The importance of virtual water at global level is likely to increase dramatically as projections made by IFPRI (Rosegrant and Ringler, 1999) show that food trade will increase rapidly: doubling for cereals and tripling for meat between 1993 and 2020. While the water and food self-sufficiency concepts sound appealing and inspire strong national feelings, these concepts often generate unrealistic perceived needs for water that are irrational and non-sustainable in most arid areas. It does require an economy that generates enough cash income from exports to cover the cost of needed food or virtual water imports (Shuval, 1998). The labour, land and capital embodied in the products must also be considered in countries where one or more of those resources are limited (Wichelns, 2001).

According to Allan and Mallat (2002), water has been regarded as a potential source of conflict and so it would have been if the governments of the Middle East states had insisted on observing their food self-sufficiency policies to the point that they had refused to import food. In practice, Middle Eastern economies have traded rationally and have gained access to very cheap water via international trade in food. Reducing the amount of water used in agriculture by importing food leaves more water for other uses. Using a lot of water just to justify a national pride of being self-sufficient in food production (especially staple foods) will then not be economical if these foods can be imported much cheaper from water rich countries (Wichelns,

2001; 2003). Nakayama (2003) suggests that existing water policies should be re-examined as aiming at food self-sufficiency by a basin country may lead to a conflict with other nations sharing an international water system. The increasing water consumption in Afghanistan is going to worsen the already deteriorated Aral Sea basin further. He recommends that the tradeoffs between real water consumption for agriculture production and virtual water consumption should be addressed from the viewpoint of security among basin countries.

Heated debates are going on worldwide in relation to the pros and cons of globalisation. The implications of trade liberalisation are also reflected in the trade of virtual water. One simple example is that subsidies in one country affect the markets in another country. If one country has high cost of water supply but low water pricing as a result of price subsidy, their export of commodities also implies the export of their subsidised water. In this case the importing country is getting subsidised water in virtual form. This can have two implications: first it will relieve the pressure on the water resources of the importing country; second it will weaken the export position of the exporting countries if they put a proper price to irrigation water.

The discourse on water, both within a single political economy and at the international level, does not include in its vocabulary the term virtual water, which however is the most significant water for many water stressed countries. It provides an extremely effective operational solution with no apparent downside (Allan, 2002). There is strong evidence that because virtual water is economically invisible and politically silent its political-economic role is not obvious. Its impacts are negative in policy terms insofar importance is given for policy reform to achieve local water use efficiency (Allan, 2001a).

Virtual water trade between nations and even continents could thus ideally be used as an instrument to improve global water use efficiency, to achieve water security in water-poor regions of the world and to alleviate the constraints on the environment by using the best suited production sites (Turton, 1999). The economic experience of the Jordan Basin has been a spectacular demonstration that natural resources such as water do not determine socio-economic development; on the contrary, socio-economic development determines water management options (Allan, 2001a).

Saving water with trade of products

A country can either produce all of its consumption goods and services domestically (using its domestic water resources) or import part of them from other countries. The decision to import from other countries can save national water resources. If the products are imported from regions that have higher water productivity than in the country of import, the total consumption of water from the global resources is smaller than if the products were produced locally. This saves water globally. However, the opposite is also true, i.e. there is a global loss, if the products are traded from less productive sites to more productive sites. These savings or losses may not always exert positive or negative impact as the terms suggest. A global water saving is good if blue water is saved in a water-poor country, using less blue water for the same output in a water-rich country. But a water saving is not good in the (more exceptional) case that blue water is saved in a water-rich country as a result of import from a highly water efficient but also highly water-scarce country.

In Figure 1.2, two possible scenarios are shown to meet the domestic demands for water intensive goods and services in a country. In scenario 1 the demand for the consumption goods is met through domestic production using the domestic water resources. Suppose, it uses 12 units of blue water (surface and ground water) and 6 units of green water (effective use of rainfall). However, in scenario 2 the country decides to import half of its consumption goods from another country where this volume is produced using only 8 units of water (4 units of blue water and 4 units of green water). In the second scenario, the total consumption of global resources is only 17 units compared to the 18 units in the first scenario. Hence, there is global water savings of 1 unit for the same units of consumption goods and services. However, if we compare the use of global blue water resources there is net saving of 2 units of blue water. This extra unit of saving in global blue water resources is at the cost of green water lost as a result of this trade.

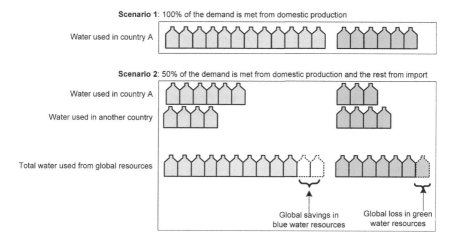

Figure 1.2. Globalisation of water as a result of international trade. In this diagram, country A either produces all its consumption goods using domestic resources (scenario I) or produces half of the demand using its own resources and imports the rest (scenario II). Here, the colour of the vessel represents the colour of water used to produced these goods.

A rapid first assessment of global water saving as a result of food trade shows a net saving of 455 Gm^3/yr (Oki and Kanae, 2004; Oki *et al.*, 2003). De Fraiture *et al.* (2004) have estimated that without cereal trade, irrigation water use would have been higher by 112 Gm^3/yr. They also come with the conclusion that only 25% of the cereal trade is water related and this may rise to 38% in the year 2025. They argue that the potential of saving should not be overestimated as water scarcity induced trade is limited, mass trade takes place within water abundant regions, saving has a meaningful contribution only if available for better alternative uses. Detailed analysis of these consequences is still lacking.

Measuring water demand and scarcity

According to Hoekstra (1998), there are three extreme view points on water scarcity. For a given water demand and if we assume that we need to have sufficient supply it is a supply-problem. Another view accepts the fact that the potential water supply is limited and demand can not go on increasing, thus making it a demand-problem. A third view on water scarcity is an economic one. This school of thought believes that if the pricing mechanism functions well, factors such as droughts, pollution and increasing demand will automatically and properly be accounted for in the water costs and thus optimise water use and allocation.

There are a number of ways in which water scarcity can be expressed. One of the most common indicators of water scarcity is the ratio between water demand in a certain area and total runoff in that area. Different terminologies are used to define it, for example: water utilisation level (Falkenmark, 1989; Falkenmark *et al.*, 1989), the use-availability ratio (Kulshreshtha, 1993), use-to-resource ratio (Raskin *et al.*, 1995). Another way of measuring scarcity is to take the ratio of the population of an area to the total runoff in that area, or the inverse of that. As the temporal and spatial variability of supply and demand are large, the numbers obtained by using available resources and population can not be compared easily among different locations.

According to Savenije (2000) the existing numbers showing the water scarcity or water availability per capita are deceptive in the sense that the earlier studies did not incorporate the available soil moisture (green water) into their available water resources. Shuval (1998) raised similar questions regarding different water stress indices suggested by Falkenmark. The 1000 m^3 per capita per year benchmark level assumes that major amounts of water must be used for agriculture and food production. It is a serious error to imply or suggest that each country can, should, or must have at its disposal enough water to be self sufficient in agricultural food production. This can and has led to irrational and often dangerous perceptions and demands concerning national water needs. Shuval (1998), here, points out this as the fundamental fallacy of the Falkenmark's water stress index or World Bank's water benchmark.

The attempts to measure and express the degree of scarcity so far are concerned with the volume of blue water withdrawn or consumed in a country or a region with respect to the volume of renewable blue water available in that country or region. In this respect it is partial as it does not include green water use in a country and the effect of import or export of water intensive goods. If we return to the basic question, why one needs to measure scarcity, one might like to see the relation between the total volume of water necessary to fulfil the entire consumption if produced domestically and the available volume of water in a country or a region. If all the consumption goods are imported then the water use in the country will be very low, so that the traditional water scarcity indicators, calculated as the ratio of withdrawal to available renewable resources, will show a very low scarcity. This gives the pseudo impression of no stress. However, the stress has been mitigated with virtual water imports. The traditional approach to express scarcity of water resources is useful only if all the consumption goods are being supplied from within a system. In the context of globalisation this is no longer a valid assumption. And, hence, one must include the virtual water flows (as a result of export and import of water intensive products) in assessing the degree of scarcity in a region or a country. The current practice of measuring demand as the withdrawal from the source should be revisited as not all the withdrawal is consumed and it does not reflect the true demand as part of the demand is already being fulfilled from imported products.

Indicators of resource utilisation

Water footprint

The water footprint concept was introduced by Hoekstra in 2002 in order to have a consumption-based indicator of water use that could provide useful information in addition to the traditional production-sector-based indicators of water use. Databases on water use traditionally show three columns of water use: water withdrawals in the domestic, agricultural and industrial sector respectively. A water expert being asked to assess the water demand in a particular country will generally add the water withdrawals for the different sectors of the economy. Although useful information, this does not tell much about the water actually needed by the people in the country in relation to their consumption pattern. The fact is that many goods consumed by the inhabitants of a country are produced in other countries, which means that it can happen that the real water demand of a population is much higher than the national water withdrawals do suggest. The reverse can be the case as well: national water withdrawals are substantial, but a large amount of the products are being exported for consumption elsewhere.

The water footprint has been developed in analogy to the ecological footprint concept as was introduced in the second half of the 1990s (Wackernagel and Jonathan, 2001; Wackernagel *et al.*, 1997; Wackernagel and Rees, 1996) The 'ecological footprint' of a population represents the area of productive land and aquatic ecosystems required to produce the resources used, and to assimilate the wastes produced, by a certain population at a specified material standard of living, wherever on earth that land may be located. Whereas the 'ecological footprint' thus shows the *area* needed to sustain people's living, the 'water footprint' indicates the *annual water volume* required to sustain a population.

The first assessment of water footprints of nations was carried out by Hoekstra and Hung (2002). They used the volume of water withdrawal (blue water) in a nation and net virtual water import to calculate the average water footprint of a nation. A more extended assessment was done by Chapagain and Hoekstra (2003a). These studies did not account for the volume of water used from green water resources for the consumption of goods produced domestically. We can now easily say that the previous studies should be considered as rudimentary. With the refinements in methodologies and concept, the water footprint can be used as an indicator to show the impact of local consumption on global water resources, both quantitatively and qualitatively.

Herendeen (2000) believes that a region's dependence now has a global dimension, and the use of an indicator like the ecological footprint can be a tool that makes the dependence visible and motivates positive response. With the examples of the Netherlands and Japan, which are often held up as economic success stories and examples for the developing world to follow, Rees (1996) pointed out that both countries enjoy high material standards at the cost of unaccounted, ecological deficits in the rest of the world.

The ecological footprint was originally conceived as a planning tool, which presents the consumption of different populations in terms of a single-point indicator describing appropriated carrying capacity, so that these populations can be compared in terms of sustainability. However, one should be careful in using the global world-average productivity as much information about impacts on regional ecosystems is lost (Lenzen and Murray, 2001). Similar views are presented by Opschoor (2000) in recommending the Dutch Minister of the Environment about whether the ecological footprint could be used as guideline to achieving sustainability. He argued that trade-related redistributions of accumulated environmental pressure should be incorporated in indicators only when (and if) sufficient empirical data are available. He further suggested that if such indicators are to relate to issues of global sustainability, they should be corrected for differences in degree of sustainability of the specific use of natural assets in situ so that the vulnerability of the local systems and their productivity in terms of environmental services are taken into account. Senbel et al. (2003) concluded that the greatest power of the ecological footprint as an indicator is not its precision or its ability to give clear direction, but rather in its conceptual simplicity and ease of communication.

Van den Bergh et al. (1999) do not think that the ecological footprint as it is presently constructed can serve as an indicator for assessing regional sustainability. They argued that, firstly, it is well-known that the human species threatens the environment, nature and biodiversity, and exhausts many natural resources. Secondly, they believe that the application of the ecological footprint on a regional level provides information that is easily misinterpreted. They listed reasons for their argument as: the ecological footprint is too aggregate, uses a fixed sustainable energy scenario, represents hypothetical rather than actual land use, makes no distinction between sustainable and unsustainable land use, does not recognize advantages of spatial concentration and specialization, and is in certain applications, biased against trade. In contrast they strongly believe that trade can in principle spatially distribute the environmental burden among the least sensitive natural systems, a point which does not seem to have attracted much attention in the literature so far. The main purpose of the ecological footprint is to raise public awareness and call people to effective political action (Cornelis et al., 2000).

The problem of data aggregation as pointed out by many authors (Cornelis et al., 2000; Lenzen and Murray, 2001; Opschoor, 2000; Senbel et al., 2003; Van den Bergh and Verbruggen, 1999) is the weakest side of the ecological footprint as a measure of sustainability. In this respect, the water footprint as an indicator is by its very nature limited to a certain domain and can be more precisely analysed. Unlike the emission of gases and its global mixing effect, the polluted water is more localised in its impact. The impact of having a larger or smaller water footprint by an individual or a nation can only be felt directly at locations either connected by global trade of products or direct sharing of water resources. In this sense, the water footprint can measure the impact more precisely. This particular aspect of the water footprint is useful in negotiating trade and water relation between different parties, even with limited global consensus.

Water dependency
With globalisation comes the risk of being dependent on other nations. It is clear that there is risk of being used for political ransom and the risk of loosing an unique identity. Basically, it is easier to meet national water deficits via the importation of water rich cereals, but this raises a series of downstream political issues that are not yet fully understood (Turton, 2002). Turton concluded that post-colonial dependency is politically risky for SADC (Southern African Development

Community) and a balance needs to be struck between a policy of national self-sufficiency and food security. The study on virtual water trade in the SADC region (Earle, 2001) shows how South Africa and Botswana rely most on virtual water imports. As their water resources have become increasingly limited they have started using water where it will receive maximum returns.

At the global level, virtual water trade has geopolitical implications: it induces dependencies between countries; it is influenced by and has implications on the world food prices as well as on the global trade negotiations and agreements. These implications are politically sensitive since it is well documented that the current low food prices of the global market are closely related to the high level of subsidies in many exporting countries and since they have a detrimental effect on the agriculture development of the countries importing food products.

Warner (2003) has raised issues like whose security is being served and who will be responsible for the vulnerability of the international trade market (e.g. price shocks as a result of unsustainable trade distortion)? He states that interdependence thus means opportunity for some, but dependence and vulnerability for others. Virtual water trade frees water-poor states from their dependency on their limited resources, but can usher in dependency of another type: dependency on unequal terms of world trade.

During the last decade, there has been a growing awareness among governments and international organisations that traditional regimes of water use are not well-adapted to major changes such as the opening up of markets, globalisation, the increasing role of the private sector, and increased consciousness of the values of environmental services. This has led to a re-visiting of national policies such as being self-sufficient. For policy analysis, it became more and more important to measure the degree of dependence on external water resources to meet domestic demands. Hoekstra and Hung (2002) proposed an indicator of water dependency as the ratio of net virtual water import to total national water appropriation. Chapagain and Hoekstra (2004) re-defines water as the ratio of the external water footprint to the total water footprint of a nation. If a country has no external water footprint, the virtual water dependency is zero, meaning that it is hundred percent self-sufficient in water for producing goods for its domestic consumption.

Local consumption global impact

Consumers are the ultimate driving force for the production of a particular product. However, the relation between a consumer and a producer is changing with time. Before the emergence of the industrial society consumers were literally the producers of the agricultural products. With industrialisation the distance between the consumer and producer increased and there were many producers and consumers meeting at local markets. Today, the consumer is situated far away from the producer and contacts are made via the supermarket and food processing industry (SIWI et al., 2005).

Keeping this distance between a producer and a consumer in mind, a consumer is largely ignorant and misinformed about the implications of the production processes on the water and environment. Hence, for using the scarce resource in a sustainable way, one needs to relate the consumption base to the production base with transparent information regarding the resource utilisation and associated environmental impacts (SIWI et al., 2005).

The consumption of goods and services creates stress on the water resources at production sites. However, the relation between the use and stress can be entirely different per location. The effect of local consumption on the water resources of other countries can be quantitatively analysed in two ways. First one can look at the absolute volume of water imported (size of the external water footprint) and the kind of virtual water imported (quality of the footprint). Second one can consider the relative volume of water imported compared to the available resources in the exporting countries. Though the size of the external water footprint can be large, it will exert less pressure in the virtual water exporting countries if the kind of water used is abundantly available in those countries (e.g. export of rain-fed maize from the USA).

Objectives

The objective of this study is to analyse opportunities and threats of virtual water trade in the context of solving national or regional problems of water shortages. A number of questions need to be answered. What are the fluxes of virtual water related to the international trade of products? Is the import of virtual water a solution to water-scarce nations or merely a threat of becoming water dependent? Can the international trade of products be a tool to enhance water use efficiency globally, or, is it a way of shifting the environmental burdens to a distant location?

The pros and cons of import and export of virtual water have been analysed from both a national and global water resources perspective. To understand the global component of fresh water demand and supply, a set of indicators have been developed. The analytical framework thus developed has been applied to different case studies.

Structure of the dissertation

The thesis is choreographed under five different themes: 'motivation and concept', 'globalisation of water resources', 'impact of local consumption on global water resources', 'case studies' and ' evaluation and outlook'. These themes are expanded into nine chapters. Figure 1.4 gives an overview of the approach and layout of the dissertation. The preceding sections have set the research objectives with the review of the literature on the concept of virtual water, water use efficiency, water scarcity, indicators for the sustainable use of resources in the context of globalisation of trade. Chapter 2 describes in detail the definitions and methodologies followed in the study.

Chapter 3 presents how international trade of goods and services indirectly connects the water resources of the trading partners. Based on the export and import of virtual water, it is shown how a country's freshwater demand is being met with global water resources. The first innovative step in this dissertation is to develop a method of calculating the virtual water content of products based on the 'product tree'.

Chapter 4 shows the consequences of international virtual water flows on the national and global water resources. It analyses the trade flows based on their respective water productivity and produces insight on whether the trade is saving the global water resources or not. It emphasises the importance of analysing the volume of water being saved in the context of the marginal utility gained. The chapter presents another innovative step by showing how global blue water resources (surface and ground water) can be saved by using the virtual water flows.

Chapter 5 translates the national consumption of agricultural and industrial goods into an equivalent water volume that has been consumed in the production process from the global water resources and assesses the water footprint of nations. The different variables determining the size of the national water footprint are analysed. This chapter presents the third innovative step by introducing the measure of water scarcity from consumption perspective. The calculation of the water footprint of a nation in this chapter is based on the top-down approach. However, it still does not include the impact of pollution from the agricultural and industrial production processes.

In Chapter 6 the virtual water flows between different regions of China are analysed and compared with the real water transfer projects being realised in China.

Chapter 7 analyses the water footprint of coffee and tea, with a case study for the Netherlands. It shows how drinking a cup of coffee or tea in the Netherlands consumes water in the coffee or tea producing countries. The coffee and tea consumption is mostly at the cost of green water (effective use of rainfall for the production of crop) which has little opportunity cost.

So far the impact of local consumption on the global water resources has been studied in terms of use of the global green and blue water resources. Chapter 8 is another case study where the impact of consumption of a cotton product is traced back to the location where it is being produced. In this case, the impact on water resources is made on the basis of type of water consumed such as blue water (irrigation), green water (effective use of rainfall) and the volume of water required to dilute the pollution from the return flows from the agricultural fields and the processing industries. The concept is further illustrated with three consumers (USA, EU 25 and Japan) and their impacts (green, blue and dilution volumes of water) on the global water resources.

Chapter 9 is the final chapter which draws from the inferences from the previous chapters. The major findings are discussed and relevant policy intervention necessary are suggested.

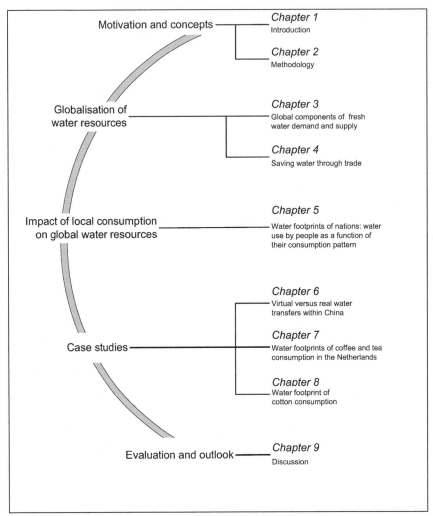

Figure 1.3. The general outline of the thesis.

Chapter 2

Methodology

The overall scheme adopted for the quantification of water footprints and other indicators is shown in Figure 2.1. First the virtual water flux (Λ) between nations is quantified based on trade volumes (T) and the associated virtual water content of the products (V). Each production region has a unique virtual water content of a product as the latter highly depends upon the agro-climatic factors at the production sites. Moreover, the processing techniques and the volume of water consumed in the processes and the processed output per unit of product processed can be different per production region even for the same product. This creates a difference in water productivity (production per unit of water consumed) between different production sites. The import of a product from highly water productive site to one with low productivity can save use of global water resources (ΔS_g) due to this difference in water productivity. The option of import of a product instead of producing itself is directly releasing the water resources (ΔS_n) for other uses at the import-sites.

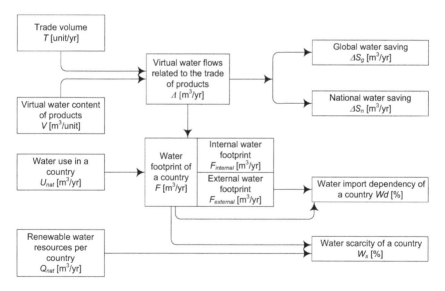

Figure 2.1 Methodological framework of analysis.

The volume of water use in each country (U_{nat}) is estimated based on the domestic production of goods and their virtual water content. Form this volume, the virtual water export related to the export of domestic production is deducted to get the volume of water consumed by the inhabitants of a country from their own domestic water resources ($F_{internal}$). Similarly, as not all the products imported are consumed domestically, the virtual water export related to the export of the imported products is deducted to get the actual volume of virtual water imported for the consumption in a country ($F_{external}$). The volume of renewable water resources available per country and their consumption of water, both from internal and external sources, can be used to get new estimates of water scarcity (W_s), water dependency (W_d) and water self sufficiency (W_{ss}). These steps are elaborated in the following sections.

Virtual water content

The virtual water content of a product is the volume of water used to produce the product, measured at the place where the product was actually produced (production site specific definition). The virtual water content of a product can also be defined as the volume of water that would have been required to produce the product in the place where the product is consumed (consumption site specific definition). In this thesis, unless otherwise mentioned explicitly, the production site-specific definition has been used. The adjective 'virtual' refers to the fact that most of the water used to produce a product is in the end not contained in the product. The real water content of products is generally negligible if compared to the virtual water content.

The virtual water content of a product p in a country (m³/unit) is calculated as the ratio of total volume of water used (U in m³) for Y unit of the production in that country.

$$V = \frac{U}{Y} \tag{1}$$

Depending the source of water used the definition virtual water content can be further elaborated as the green virtual water content (V_g, resulting from the use of effective rainfall applicable for crop production), blue virtual water content (V_b, resulting from surface and renewable ground water sources), fossil virtual water content (from mining the fossil ground water resources) etc. One can even translate the pollution effect of a production system into the equivalent volume water necessary for dilution per unit of production.

$$V_g = \frac{I_e}{Y} \tag{2}$$

$$V_b = \frac{U_b}{Y} \tag{3}$$

$$V_f = \frac{U_f}{Y} \tag{4}$$

where I_e is the effective rainfall in crop production, U_b is the volume of blue water used for the production, and U_f is the volume of water use from mining the fossil ground water resources.

Virtual water content of primary crops

The virtual water content of a crop c (m^3/ton) is calculated as the ratio of total volume of water used for crop production, U_c (m^3), to the total volume of crop produced, Y_c (ton).

$$V_c = \frac{U_c}{Y_c} \tag{5}$$

The average virtual water content of a crop c in a country, $V_{c,n}$ (m^3/ton), is calculated as the ratio of total volume of water used for the production of crop c to the total volume of crop produced in that country.

$$V_{c,n} = \frac{\sum_{a=0}^{a=A_c} U_c}{\sum_{a=0}^{a=A_c} Y_c} \tag{6}$$

where A_c is the total harvest area (ha) of crop c in the country. The total volume of water used for crop production U_c, is calculated as:

$$U_c = R_c \times A_c \tag{7}$$

where R_c is the crop water requirement (m^3/ha) for the entire growth period of a crop c. In this thesis, it is assumed that crop water requirement is fully met either by irrigation or by rainfall, which is calculated as:

$$R_c = 10000 \times \sum_{d=1}^{d=l_p} E_c \tag{8}$$

where l_p is the length of growing period of crop (days), E_c is the total evaporation from crop field (m/day). The crop water requirement of rice cannot be calculated directly using Equation 8. In addition to evaporation from the paddy field, there is a considerable amount of percolation from the field, which varies with the soil type and ground water table at the farm. Assuming that rice is normally grown in a loam and loamy clay, we have added 300 mm of water for percolation during plantation period. Although the percolated water is available further downstream, in this thesis it is assumed to be polluted enough for direct uses downstream.

The evaporation from a crop field E_c (m/day) is calculated using the crop coefficient (K_c) for the respective growth period.

$$E_c = K_c \times E_r \tag{9}$$

where E_r is the reference crop evaporation (m/day) at that particular location and time. The steps in the calculation of virtual water content of a crop are shown in Figure 2.2.

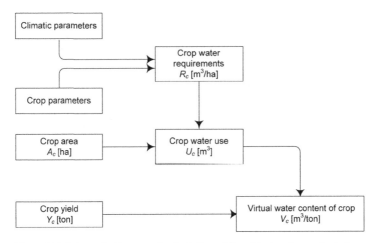

Figure 2.2. Schematic diagram of calculation of the virtual water content of a crop.

Reference crop evaporation
The reference crop evaporation (E_r) is the evaporation rate from a reference surface, not short of water. The Penman-Monteith equation is used to estimate the reference evaporation E_r. The FAO Expert Consultation on Revision of FAO Methodologies for Crop Water Requirements accepted the following unambiguous definition for the reference surface: "A hypothetical reference crop with an assumed crop height of 0.12 m, a fixed surface resistance (r_s) of 70 (s/m) and an albedo of 0.23" (Allen *et al.*, 1998). The reference surface closely resembles an extensive surface of green grass of uniform height, actively growing, completely shading the ground and with adequate water. The requirements that the grass surface should be extensive and uniform result from the assumption that all fluxes are one-dimensional upwards. The only factors affecting E_r are climatic parameters. E_r expresses the evaporating power of the atmosphere at a specific location and time of the year and does not consider the crop characteristics and soil factors.

The classical Penman-Monteith equation to estimate evaporation is:

$$E = \frac{\Delta(R_n - G) + c_p \rho_a \dfrac{(e_s - e_a)}{r_a}}{\left[\Delta + \gamma\left(1 + \dfrac{r_s}{r_a}\right)\right]\rho_w \lambda} \tag{10}$$

Where
E	:	evaporation [m/d],
λ	:	latent heat of vaporization, 2.45 [MJ/kg],
Δ	:	slope of the vapour pressure curve [kPa/°C] (Equation 13),
c_p	:	specific heat at constant pressure, 1.013×10^{-3} [MJ/kg/°C],
ρ_a	:	mean air density at constant pressure, [kg/m³], (Equation 11)
ρ_w	:	density of water, taken equal to 1000, [kg/m³], (Equation 11)
γ	:	psychrometric constant [kPa/°C] (Equation 15),
e_s	:	saturation vapour pressure [kPa] (Equation 17),
R_n	:	net radiation at the crop surface [MJ/m²/day] (Equation 19),

G :   soil heat flux [MJ/m^2/day] (Equation 29),
r_a :   aerodynamic resistance[d/m],
r_s :   bulk surface resistance of the crop canopy and soil [d/m],
e_a :   actual vapour pressure [kPa],
e_s-e_a :   vapour pressure deficit [kPa].

For reference surface, the bulk surface resistance of the crop canopy and soil is taken equal to 70 s/m (Allen *et al.*, 1998). The mean air density at constant pressure (ρ_a) and the aerodynamic resistance (r_a) are calculated using Equation 12 and 12 respectively.

$$\rho_a = \frac{P}{1.01(T+273)R} \tag{11}$$

$$r_a = \frac{208}{U_2} \tag{12}$$

Where

U_2 :   wind speed measured at 2 m height [m/s],
P :   atmospheric pressure [kPa] (Equation 16),
T :   average air temperature [°C] (Equation 14),

Equation 10 is applied with a time step of a month. For all input data, monthly averages have been taken. A smooth graph of E_r over the year has been obtained by assuming that the calculated monthly averages hold for the 15th of the month and by assuming linear development in between the 15th of one month and 15th of next month. The various parameters in Equation 10 are calculated in different steps.

The slope of saturation vapour pressure curve (Δ) at air temperature T (kPa/°C) is calculated as:

$$\Delta = \frac{4098\left[0.6108 \times e^{\left(\frac{17.27T}{T+237.3}\right)}\right]}{(T+237.3)^2} \tag{13}$$

where T [°C] is air temperature (Equation 14). The slope of the vapour pressure curve is calculated using mean air temperature T_{mean}, calculated as an average of T_{max} (daily maximum temperature) and T_{min} (daily minimum temperature).

$$T_{mean} = \frac{T_{max} + T_{min}}{2} \tag{14}$$

The psychrometric constant, γ (kPa/°C) is calculated as:

$$\gamma = \frac{c_p P}{\varepsilon \lambda} = 0.665 \times 10^{-3} P \tag{15}$$

Where

P :   atmospheric pressure [kPa] (Equation 16),
λ :   latent heat of vaporization, 2.45 [MJ/kg],
c_p :   specific heat at constant pressure, 1.013x10^{-3} [MJ/kg/°C],
ε :   ratio molecular weight of water vapour/dry air = 0.622.

The specific heat at constant pressure is the amount of energy required to increase the temperature of a unit mass of air by one degree at constant pressure. For average atmospheric conditions a value c_p = 1.013×10^{-3} (MJ/kg/°C) can be used. Atmospheric pressure P in kPa for a location (at an elevation of z m above mean sea level) is calculated as follows:

$$P = 101.3 \left(\frac{293 - 0.0065z}{293} \right)^{5.26} \tag{16}$$

Mean saturation vapour pressure e_s is calculated as:

$$e_s = \frac{e^0_{(T_{max})} + e^0_{(T_{min})}}{2} \tag{17}$$

where $e^0_{(Tmax)}$ and $e^0_{(Tmin)}$ are calculated as:

$$e^0_{(T)} = 0.6108 \times \exp\left(\frac{17.27T}{T + 237.3} \right) \tag{18}$$

The net radiation is the difference between the incoming net shortwave radiation (R_{ns}) and the outgoing net longwave radiation (R_{nl}):

$$R_n = R_{ns} - R_{nl} \tag{19}$$

The net shortwave radiation (R_{ns}) resulting from the balance between incoming and reflected solar radiation is given by:

$$R_{ns} = (1 - \alpha) R_s \tag{20}$$

Where
  R_{ns} :   net solar or shortwave radiation [MJ/m²/day]
  α :   albedo or canopy reflection coefficient, which is 0.23 for the hypothetical grass reference crop,
  R_s :   the incoming solar radiation [MJ/m²/day] (Equation 22).

Net longwave radiation (R_{nl}) is given by the Stefan-Boltzmann equation.

$$R_{nl} = \sigma \left[\frac{T^4_{(max,K)} + T^4_{(min,K)}}{2} \right] \left(0.34 - 0.14\sqrt{e_a} \right) \left(1.35 \frac{R_s}{R_{so}} - 0.35 \right) \tag{21}$$

Where
  R_{nl} :   net outgoing longwave radiation [MJ/m²/day],
  σ :   Stefan-Boltzmann constant [4.903×10^{-9} MJ/K⁴/m²/day],
  $T_{max, K}$:   maximum absolute temperature during the 24-hour period [K=°C+273.16],
  $T_{min, K}$:   minimum absolute temperature during the 24-hour period [K=°C+273.16],
  e_a :   actual vapour pressure [kPa],
  R_s/R_{so} :   relative shortwave radiation (limited to ≤ 1.0),
  R_s :   solar radiation [MJ/m²/day] (Equation 22),
  R_{so} :   clear-sky radiation [MJ/m²/day] (Equation 23).

Solar radiation (R_s) can be calculated with the Angstrom formula, which relates solar radiation to extraterrestrial radiation and relative sunshine duration:

$$R_s = \left(a_s + b_s \frac{n}{N} \right) R_a \tag{22}$$

Where
R_s	:	solar or shortwave radiation [MJ/m^2/day],
n	:	actual duration of sunshine [hour],
N	:	maximum possible duration of sunshine or daylight hours [hour],
R_a	:	extraterrestrial radiation [MJ/m^2/day] (Equation 24),
a_s	:	regression constant, expressing the fraction of extraterrestrial radiation reaching the earth on overcast days (n = 0),
a_s+b_s	:	fraction of extraterrestrial radiation reaching the earth on clear days (when n = N).

Depending on atmospheric conditions (humidity, dust) and solar declination (latitude and month), the Angstrom values as and bs will vary. Where no actual solar radiation data are available and no calibration has been carried out for improved as and bs parameters, the values as = 0.25 and bs = 0.50 are taken as recommended by (Allen et al., 1998). The clear-sky radiation, R_{so}, when n = N, is calculated as:

$$R_{so} = \left(0.75 + 2 \times 10^{-6} z \right) R_a \tag{23}$$

where, R_a is extraterrestrial radiation (MJ/m^2/day, Equation 24) and z is the elevation above mean sea level (m). The extraterrestrial radiation, Ra, for each day of the year and for different latitudes can be estimated from the solar constant, the solar declination and the time of the year.

$$R_a = \frac{24 \times (60)}{\pi} G_{sc} d_r \left[\omega_s \sin(\varphi) \sin(\delta) + \cos(\varphi) \cos(\delta) \sin(\omega_s) \right] \tag{24}$$

Where
R_a	:	extraterrestrial radiation [MJ/m^2/day],
G_{sc}	:	solar constant = 0.0820 [MJ/m^2/day],
d_r	:	inverse relative distance Earth-Sun (Equation 25),
ω_s	:	sunset hour angle [rad] (Equation 28),
φ	:	latitude [rad] (Equation 27),
δ	:	solar decimation [rad] (Equation 26).

Here d_r and δ are calculated as:

$$d_r = 1 + 0.033 \cos\left(\frac{2\pi}{365} J \right) \tag{25}$$

$$\delta = 0.409 \sin\left(\frac{2\pi}{365} J - 1.39 \right) \tag{26}$$

Here J is the number of the day in the year between 1 (1 January) and 365 or 366 (31 December). The latitude (φ) expressed in radians is positive for the northern hemisphere and negative for the southern hemisphere.

$$\varphi[\text{radians}] = \frac{\pi}{180}[\text{decimal degrees}] \tag{27}$$

The sunset hour angle (ωs) is given by:

$$\omega_s = \arccos[-\tan(\varphi)\tan(\delta)] \tag{28}$$

Complex models are available to describe soil heat flux. Because soil heat flux is small compared to R_n, particularly when the surface is covered by vegetation, for monthly average G, we can use the following:

$$G_{month,i} = 0.07\left(T_{month,i+1} - T_{month,i-1}\right) \tag{29}$$

Where

$T_{month,\,i}$	:	mean air temperature of month i [°C] (Equation 14),
$T_{month,\,i-1}$	:	mean air temperature of previous month [°C] (Equation 14),
$T_{month,\,i+1}$	:	mean air temperature of next month [°C] (Equation 14).

Crop coefficients
The crop evaporation (E_c) differs distinctly from the reference evaporation (E_r), as the ground cover, canopy properties and aerodynamic resistance of the crop are different from grass. The effects of characteristics that distinguish field crops from grass are integrated into the crop coefficient (K_c).

The major factors determining K_c are crop variety, climate and crop growth stage. For instance, more arid climates and conditions of greater wind speed will have higher values for K_c. More humid climates and conditions of lower wind speed will have lower values for K_c. As the crop develops, the ground cover, crop height and the leaf area change. Due to differences in evaporation during the various growth stages, the K_c for a given crop will vary over the growing period. The growing period can be divided into four distinct growth stages: initial, crop development, mid-season and late season.

In general the K_c of a crop varies along with the stages. The general character of a K_c curve is shown in Figure 2.3. The *initial stage* runs from planting date to approximately 10% ground cover. The length of the initial period is highly dependent on the crop, the crop variety, the planting date and the climate. For perennial crops, the planting date is replaced by the 'green up' date, i.e., the time when the initiation of new leaves occurs. During the initial period, the leaf area is small, and evaporation is predominately in the form of soil evaporation. Therefore, the K_c during the initial period is large when the soil is wet from irrigation and rainfall and is low when the soil surface is dry.

The *crop development stage* runs from 10% ground cover to effective full cover, which for many crops occurs at the initiation of flowering. As the crop develops and shades more and more of the ground, evaporation becomes more restricted and transpiration gradually becomes the major process. During the crop development stage, the K_c value corresponds to the extent of ground cover. Typically, if the soil surface is dry, $K_c = 0.5$ corresponds to about 25-40% of the ground surface covered by vegetation. A K_c value of 0.7 often corresponds to about 40-60% ground cover.

These values will vary, depending on the crop, frequency of wetting and whether the crop uses more water than the reference crop at full ground cover.

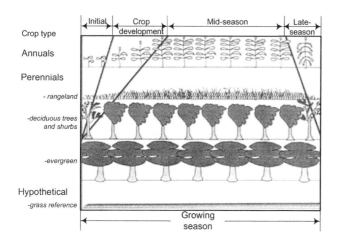

Figure 2.3. Crop growth stages for different types of crops (Allen *et al.*, 1998).

The *mid-season stage* runs from effective full cover to the start of maturity. The start of maturity is often indicated by the beginning of the ageing, yellowing or senescence of leaves, leaf drop, or the browning of fruit to the degree that the crop evaporation is reduced relative to the reference crop evaporation *Er*. The mid-season stage is the longest stage for perennials and for many annuals, but it may be relatively short for vegetable crops that are harvested fresh for their green vegetation. In the mid-season stage K_c has its maximum value and remains constant. Deviation of K_c from the reference value '1' is primarily due to differences in crop height and resistance between the grass reference surface and the actual crop surface.

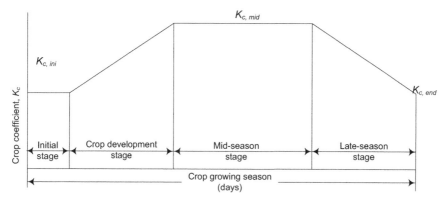

Figure 2.4. Development of K_c during the crop growing season.

The *late season* stage runs from the start of maturity to harvest or full senescence. The calculation of crop evaporation is presumed to end when the crop is harvested, dries out naturally, reaches full senescence, or experiences leaf drop. For some perennial vegetation in frost-free climates, crops may grow year round so that the date of termination may be taken the same as the date of 'planting'. The K_c value at the end of the late season stage reflects crop and water management practices. The K_c value is high if the crop is frequently irrigated until harvested fresh. If the crop is allowed to senesce and to dry out in the field before harvest, the K_c value will be small.

Virtual water content of live animals

The virtual water content of an animal at the end of its life span is defined as the total volume of water that was used to grow and process its feed, to provide its drinking water, and to clean its housing and the like. It depends on the breed of an animal, the farming system, the feed consumption and the climatic conditions of the place where the feed is grown. There are three components to the virtual water content V_a of live animal a:

$$V_a = V_{a,feed} + V_{a,drink} + V_{a,serv} \tag{30}$$

$V_{a,feed}$, $V_{a,drink}$ and $V_{a,serv}$ represent the virtual water content of animal a related to feed, drinking water and service water consumption respectively, expressed in cubic metres of water per live animal (Figure 2.5).

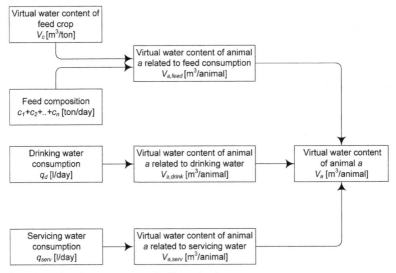

Figure 2.5. Virtual water content of a live animal.

The virtual water content of an animal at the end of its life span from the feed consumed has two parts: first is the actual water that is required to prepare the feed mix and the second is the virtual water incorporated in the various feed ingredients.

$$V_{a,feed} = \int_{birth}^{slaughter} \left\{ q_{mix} + \sum_{c=1}^{n} cV_c \right\} dt \qquad (31)$$

The variable q_{mix} represents the volume of water required for mixing the feed (m^3/day). c $[a,c]$ is the quantity of feed crop c consumed by the animal, expressed in tons per day.

The virtual water content of an animal originating from drinking is equal to the total volume of water withdrawn for drinking water supply, calculated over the entire life span of the animal.

$$V_{a,drink} = \int_{birth}^{slaughter} q_{drink} \, dt \qquad (32)$$

The virtual water content of an animal from the service water used is equal to the total volume of water used to clean the farmyard, wash the animal and other services necessary to maintain the environment during the entire life span of the animal.

$$V_{a,serv} = \int_{birth}^{slaughter} q_{serv} \, dt \qquad (33)$$

q_{drink} and q_{serv} are the daily drinking water requirement and the daily service water requirement of the animal respectively (m^3/day).

Virtual water content of processed crop and livestock products
The virtual water content of a processed product depends on the virtual water content of the primary crop or live animal from which it is derived. The virtual water content of the primary crop or live animal is distributed over the different products from that specific crop or animal. We have assumed that each individual crop or livestock product p comes from one and only one particular type of primary crop c or live animal a. For simplification it is further assumed that a product p exported from a certain country e is actually produced from a primary crop c or animal a grown within that country using the domestic resources only.

For the sake of systematic analysis we assume 'levels of production'. The products derived directly from a primary crop or a live animal are called primary products. For example, cows produce milk, a carcass and skin as their primary products. From paddy (rice) we get husked rice as a primary crop product. From soybean we get soybean crude oil and soybean oil cakes as primary crop products. Some of these primary products are further processed into so-called secondary products, such as cheese and butter made from the primary product milk, flour made from husked rice and meat and sausage processed from the carcass.

The virtual water content of a processed product from a primary crop or a live animal includes (part of) the virtual water content of the primary crop or live animal plus the processing water needed. The processing water requirement is calculated as follows:

$$R_{proc} = \frac{Q_{proc}}{\chi_{proc}} \qquad (34)$$

Here R_{proc} is the processing water requirement per ton of primary crop c or live animal a, for processing primary products in a country (m^3/ton). Q_{proc} is the total volume of processing water required (m^3) to process crop c or animal a. χ_{proc} is the total weight of the primary crop or live animal processed.

The sum of processing water requirement (R_{proc}) and the virtual water content of the primary crop (V_c) or the virtual water content of the live animal (V_a) should be attributed to the processed products in a logical way. To do this we introduce the terms *product fraction* and *value fraction*. The product fraction f_p of product p is defined as the weight of the primary product obtained per ton of primary crop or live animal. For example, if one ton of paddy (rice) produces 0.62 ton of husked rice, the product fraction of husked rice is 0.62. The f_p for a crop or a livestock product is calculated as:

$$f_p = \frac{\chi}{\chi_{proc}}$$

(35)

Here χ is the weight of primary product p obtained from processing χ_{proc} ton of primary crop c or live animal a. Generally the product fraction is less than one, because the product is derived from just part of the animal or crop. However, if a product is obtained during the lifetime of an animal, as in the case of milk and eggs, the fraction can be greater than one.

If there are more than two products obtained while processing a primary crop or a live animal, we need to distribute the virtual water content of the primary crop or the live animal to its products based on value fractions and product fractions. The value fraction, f_v, of a product is the ratio of the market value of the product to the aggregated market value of all the products obtained from the primary crop or live animal:

$$f_v = \frac{v_p \times f_p}{\sum v_p \times f_p}$$

(36)

The denominator is totalled over the primary products that originate from the primary crop c or the animal a. The variable v_p is the market value of product p (US\$/ton). Hence, the virtual water content (V_p) of primary product p in m^3/ton is:

$$V_p = \left(V + R_{proc}\right) \times \frac{f_v}{f_p}$$

(37)

In a similar way we can calculate the virtual water content for secondary and tertiary products, etc. The first step is always to obtain the virtual water content of the input (root) product and the water necessary to process it. The total of these two elements is then distributed over the various output products, based on their product fraction and value fraction.

For example, soybean can be either processed into soybean flour or into oil products (Figure 2.6). One ton of soybean produces 0.85 tons of soybean flour. If the virtual water content of soybean is 1789 m^3/ton, the virtual water content of soybean flour is 2105 (= 1789/0.85) m^3/ton. Instead of processing soybeans into soybean flour, we can also process soybeans into soybean crude oil ($f_p = 0.18$ ton per ton of soybean) and soybean oil cake ($f_p = 0.79$ ton per ton of soybean). The global average market value of soybean crude oil is 502 US\$/ton and soybean oil cake is 219 US\$/ton. The total market value of soybean crude oil is, thus, 90 US\$ (= 502*0.18) and the market value of soybean oil cake produced is 173 US\$ (=0.79*219). The total market value produced is 263 US\$ (= 90+173). Hence, the value fraction of soybean crude oil is 0.343 ($f_v = 90/263$) and for the soybean oil cake it is 0.657 ($f_v = 173/263$). Neglecting process water requirements, the virtual water content of the two products from soybean can be calculated as shown in the Figure 2.6.

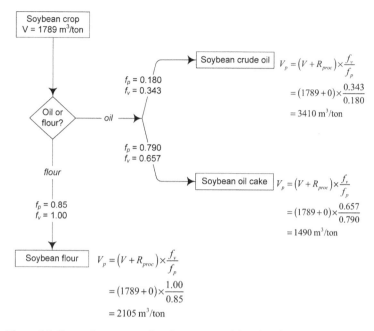

Figure 2.6. Processing one ton of soybean crop and the virtual water content of its products.

Virtual water content of industrial products
The virtual water content of an industrial product can be calculated in a similar way as described earlier for agricultural products. There are however numerous categories of industrial products with a diverse range of production methods and detailed standardised national statistics related to the production and consumption of industrial products are hard to find. As the global volume of water withdrawn in the industrial sector is only 716 Gm^3/yr ($\approx$ 10% of total global water use), in this research an average virtual water content per dollar added value in the industrial sector (V_{ind}, m^3/US$) is calculated as:

$$V_{ind} = \frac{U_{ind}}{D_{ind}} \tag{38}$$

Here U_{ind} is the industrial water use (m^3/yr) in a country, assumed to be equal to the industrial water withdrawal in a country, while D_{ind} is the added value of the industrial sector, which is one component contributing to the gross domestic product D_{gross} (US$/yr).

The global average virtual water content of industrial products ($V_{ind,g}$) is defined as:

$$V_{ind,g} = \frac{\sum\limits_{e=1}^{n} U_{ind,e}}{\sum\limits_{e=1}^{n} D_{ind,e}} \tag{39}$$

Virtual water flows

The virtual water flow between two nations or regions is the volume of virtual water that is being transferred from one place to another as a result of product trade. Virtual water flows between nations related to the trade of agricultural products have been calculated by multiplying commodity trade flows by their associated virtual water content:

$$\Lambda = TV \tag{40}$$

where Λ is the virtual water flow (m^3/yr) as a result of export of product T (ton/yr) from country exporting country e to importing country i, and V is the virtual water content of the product (m^3/ton) in the exporting country e.

The virtual water export of a country or region is the volume of virtual water associated with the export of goods or services from the country or region. It is the total volume of water required to produce the products for export. Similarly, the virtual water import of a country or region is the volume of virtual water associated with the import of goods or services into the country or region. It is the total volume of water required (in the export countries) to produce the products for import. Viewed from the perspective of the importing country, this water can be seen as an additional source of water that comes on top of the domestically available water resources.

For industrial products, the monetary value of the trade is taken as the commodity trade to calculate the virtual water flows. The total volume of virtual water exported from country e as a result of export of industrial products $\Lambda_{ind,ex}$(m^3/yr) is obtained by multiplying the export value of industrial products M_e(US\$/yr) by the virtual water content per dollar V_{ind}(m^3/US\$):

$$\Lambda_{ind,exp} = V_{ind} M_e \tag{41}$$

The virtual water import related to the import of industrial products $\Lambda_{ind,im}$(m^3/yr) is calculated using the global average virtual water content in the industrial sector $V_{ind,g}$(m^3/US\$) and the import value of the industrial products M_i(US\$/yr):

$$\Lambda_{ind,im} = V_{ind,g} M_i \tag{42}$$

The virtual water balance $W_{b,vir}$ of a country e over a certain time period is defined as the net import of virtual water over this period, which is equal to the gross import of virtual water minus the gross export:

$$W_{b,vir} = \sum_{i=1}^{n} \left(\Lambda_{im,i} - \Lambda_{ex,i} \right) \tag{43}$$

A positive virtual water balance implies net inflow of virtual water to the country from other countries. A negative balance means net outflow of virtual water.

The various steps in the calculation of virtual water flows leaving and entering a country are presented schematically in Figure 2.7.

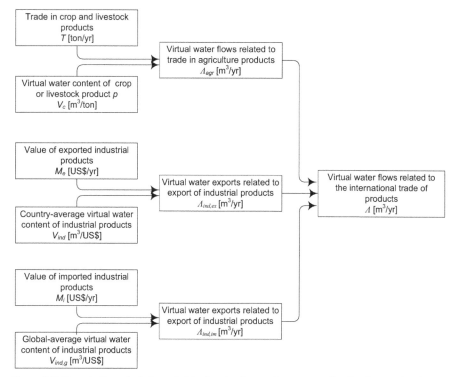

Figure 2.7. Steps in the calculation of virtual water flows of a country related to international trade of agricultural and industrial products.

Water footprints

The water footprint of an individual, business or nation is defined as the total volume of fresh water that is used to produce the foods and services consumed by the individual, business or nation. A water footprint of a nation (F) is generally expressed in terms of the volume of water use per year in a country.

$$F = U_{agr} + U_{ind} + U_{dom} + \Lambda_{im} - \Lambda_{ex} \tag{44}$$

The first three components U_{agr} (water use in agricultural productions), U_{ind} (water use in industrial products) and U_{dom} (water use for household consumption) are volume of water used from the national economy to produce goods or services either for domestic consumption or for export or both. The last two components represent the balance of virtual water in a national economy as a result of international trade in products.

The water footprint of a nation can be classified into different categories based on the origin of the water used for the production of goods and services consumed. The part of the water footprint originating from the national water resources is called the *internal water footprint* of a nation ($F_{internal}$) and the part of the water footprint originating from outside of a national territory is called *external water footprint* of a nation ($F_{external}$).

$$F = F_{internal} + F_{external} \tag{45}$$

$$F_{internal} = U_{agr} + U_{ind} + U_{dom} - \Lambda_{ex,dom} \tag{46}$$

$$F_{external} = \Lambda_{im} - \Lambda_{re-ex} \tag{47}$$

where $\Lambda_{ex,dom}$ is the volume of virtual water export to other countries insofar related to export of domestically produced products (m³/yr) and Λ_{re-ex} the volume of virtual water exported to other countries as a result of re-export of imported products (m³/yr).

If a part of the water footprint is resulting from the use of blue water resources it is called *blue water footprint*, and the part from the effective use of rainfall is called *green water footprint*. One can even define *fossil water footprint* if the goods are produced from mining the ground water. The calculation for these different kinds of footprints can be done by using the appropriate virtual water content of the products (Equations: 2, 3 and 4) in the calculation schemes. The steps in the water footprint calculation are schematically shown in Figure 2.8.

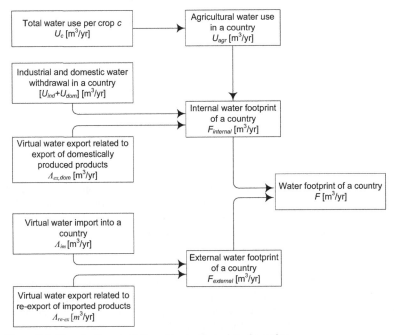

Figure 2.8. Steps in the calculation of water footprint of a nation.

For cross-country comparisons, determination of an individual water footprint is useful. The water footprint of an individual F_{indv} is defined as the total volume of water used for the production of the goods and services consumed by the individual. It can be estimated by multiplying all goods and services consumed by their respective virtual water content:

$$F_{indv} = \sum_{x=1}^{X} p_x V_p \qquad (48)$$

where p_x is the product consumed, V_p is the virtual water content of the respective product, and X is the number of products consumed. However, in the absence of data an average individual water footprint, one can estimate the per-capita water footprint of a nation (F_{pc}) by dividing the national water footprint (F) by its population (N_{pop}) as:

$$F_{pc} = \frac{F}{N_{pop}} \qquad (49)$$

Water scarcity, dependency and self sufficiency

Water scarcity has often been defined as the ratio of actual water withdrawals to the available renewable water resources. This supply-oriented definition is useful from a production perspective, but does not express the scarcity from a demand perspective. In this study, water scarcity is defined as the ratio of the total water footprint of a country or region (F) to the total renewable water resources, Q_{nat} (Equation 50). The national water scarcity can be more than 100% if a nation consumes more water than domestically available. As a measure of water availability we take here the 'total renewable water resources (actual)' as defined by FAO in their AQUASTAT database.

$$W_S = \frac{F}{Q_{nat}} \times 100 \qquad (50)$$

Countries with import of virtual water depend, de facto, on the water resources available in other parts of the world. The virtual water import dependency (W_d) of a country or region is defined as the ratio of the external water footprint of the country or region to its total water footprint.

$$W_d = \frac{F_{external}}{F} \times 100 \qquad (51)$$

Water self-sufficiency of a country (W_{ss}) is the ratio of the internal water footprint to the total water footprint of a country or region (Equation 52). It denotes the national capability of supplying the water needed for the production of the domestic demand for goods and services. Self-sufficiency is 100% if all the water needed is available and indeed taken from within the own territory. Water self-sufficiency approaches zero if the demand for goods and services in a country is largely met with virtual water imports.

$$W_{ss} = \frac{F_{internal}}{F} \times 100 \qquad (52)$$

Water savings and losses

A nation can save its domestic water resources by importing a water-intensive product rather than producing it domestically or loose its domestic water resources by exporting them. The net national saving $\Delta S_{n,i}$ (m³/yr) of a country i as a result of trade of product p is:

$$\Delta S_{n,i} = V_{p,i}\, T_{im,p,i} - V_{p,i}\, T_{ex,p,i} \tag{53}$$

Where $V_{p,i}$ is the virtual water content of product in the importing country i, $T_{imp,p,i}$ is the import of product in country i and $T_{ex,p,i}$ is the export of products from country i. The national water saving can have negative sign, which means national water loss.

International trade can save water globally if a water-intensive commodity is traded from an area where it is produced with high water productivity (resulting in products with low virtual water content) to an area with lower water productivity. The global water saving ΔS_g (m³/yr) through the trade of a product p from an exporting country e to an importing country i, is:

$$\Delta S_{g,p,e,i} = T_{p,e,i} \left(V_{p,i} - V_{p,e} \right) \tag{54}$$

where $T_{p,e,i}$ is the amount of trade (ton/yr) between the two countries, $V_{p,i}$ is the virtual water content of product in the importing country i, and $V_{p,e}$ is the virtual water content of product in the exporting country e.

Chapter 3

The global component of freshwater demand and supply[1]

In the world of today, people in Japan indirectly affect the hydrological system in the United States and people in the Netherlands indirectly impact on the regional water systems in Brazil. Much has been reported about the expected effects of past and ongoing local emissions of greenhouse gasses on the future global temperature, evaporation and precipitation patterns. Little attention has been paid however to a second mechanism through which people affect water systems in other parts of the world. This second mechanism, which is actually much more visible already today, is through global trade. International trade in agricultural and industrial commodities creates a direct link between the demand for water-intensive commodities (notably crops) in countries such as Japan and the Netherlands and the water use for production of export commodities in countries such as the United States and Brazil. The water use for producing export commodities to the global market significantly contributes to the change of regional water systems. Japanese consumers put pressure on water resources in the US, contributing to the mining of aquifers and emptying of rivers in North America. We know the examples of the mined Ogallala Aquifer and emptied Colorado River. Dutch consumers contribute to a significant degree to the water demand in Brazil.

While it is generally argued that the river basin is the appropriate unit for analyzing freshwater availability and use, this paper argues that it becomes increasingly important to put freshwater issues in a global context. Although other authors have already argued so earlier (Postel et al., 1996; Vörösmarty et al., 2000), this paper adds a new dimension to the argument. International trade of commodities implies large-distance transfers of water in virtual form, where virtual water is understood as the volume of water that is required to produce a commodity and that is thus virtually embedded in it (Allan, 1993, 1994). Knowing the virtual water flows entering and leaving a country can put a completely new light on the actual water scarcity of a country. Jordan, as an example, imports about 5 to 7 billion cubic meter of virtual water per year (Chapagain and Hoekstra, 2003a; Haddadin, 2003), which is in sheer contrast with the 1 billion cubic meter of annual water withdrawal from domestic water sources. As another example, Egypt, with water self-sufficiency high on the political agenda and with a total water withdrawal inside the country of 65 billion cubic meter per year, still has an estimated net virtual water import of 10 to 20 billion cubic meter per year (Chapagain and Hoekstra, 2003a; Yang and Zehnder, 2002; Zimmer and Renault, 2003).

In the past few years a number of studies have become available showing that the virtual water flows between nations are substantial. All studies indicate that the

---

[1] Based on: Chapagain and Hoekstra (2004); Chapagain and Hoekstra (submitted-a) as manuscript to Water International.

global sum of international virtual water flows must exceed 1000 billion cubic meters per year (Chapagain and Hoekstra, 2003a; Hoekstra and Hung, 2005; Oki *et al.*, 2003; Zimmer and Renault, 2003). The current study comes up with more accurate and comprehensive estimates of international virtual water flows in the period 1997-2001 and analyses what these virtual water flows mean in terms of global water use efficiency and water import dependency of regions.

International virtual water flows have been calculated by multiplying commodity trade flows by their associated virtual water content (Equation 40). The trade between 243 countries are taken into the calculations, for which international trade data are available in the Personal Computer Trade Analysis System of the International Trade Centre. It covers trade data from 146 reporting countries disaggregated by product and partner countries (ITC, 2004). The calculations have been carried out for 285 crop products and 123 livestock products.

The virtual water content of primary crops has been calculated based on crop water requirements and yields. Crop water requirement were calculated per crop and per country using the methodology developed by FAO (Allen *et al.*, 1998). The virtual water content of crop products has been calculated based on product fractions (ton of crop product obtained per ton of primary crop) and value fractions (the market value of one crop product divided by the aggregated market value of all crop products derived from one primary crop). The virtual water content of live animals was calculated based on the virtual water content of their feed and the volumes of drinking and service water consumed during their lifetime. Eight major animal categories were included in the study: beef cattle, dairy cows, swine, sheep, goats, fowls/poultry (meat purpose), laying hens and horses. The calculation of the virtual water content of livestock products has again been based on product fractions and value fractions, following the methodology described in Chapter 2.

Data on trade in industrial products have been taken from the World Trade Organization (WTO, 2004). Virtual water imports and exports have been calculated by multiplying monetary data on international trade of industrial products by country specific data on the average virtual water content per dollar of industrial products. In this approach, all industrial products are included implicitly.

International virtual water flows

The calculations show that the global virtual water flows during the period 1997-2001 added up to an average of 1625 Gm^3/yr. The major share (61%) of the virtual water flows between countries is related to international trade of crops and crop products. Trade in livestock products contributes 17% and trade in industrial products 22%. The total volume of international virtual water flows includes virtual water flows that are related to re-export of imported products. The global volume of virtual water flows related to export of domestically produced products is 1197 Gm^3/yr (Table 3.1 and 3.2). With a total global water use of 7451 Gm^3/yr, this means that 16% of the global water use is not meant for domestic consumption but for export. In the agricultural sector, 15% of the water use is for producing export products; in the industrial sector this is 34%.

Table 3.1. International virtual water flows per sector. Period 1997-2001.

	Gross virtual water flows related to the international trade of products (Gm³/yr)		
	Agricultural	Industrial	Total
Virtual water export related to export of domestically produced products	957	240	1197
Virtual water export related to re-export of imported products	306	122	428
Total virtual water export	1263	362	1625

Table 3.2. Global water use per sector. Period 1997-2001.

	Agricultural sector	Industrial sector	Domestic sector	Total
Global water use (Gm³/yr)	6391	716	344	7451
Water use in the world not used for domestic consumption but for export (%)	15	34	0	16

The major water exporters are the US, Canada, France, Australia, China, Germany, Brazil, the Netherlands and Argentina. The major water importers are the US, Germany, Japan, Italy, France, the Netherlands, the UK and China. Table 3.3 presents the virtual water flows for a number of selected countries. Import of water in virtual form can substantially contribute to the total 'water supply' of a country. The Netherlands imports for instance a net amount of (virtual) water equivalent to the annual net precipitation in the country. Jordan imports a volume of water in virtual form equivalent to *five times* its own annual renewable water resources.

A national virtual water flow balance can be drafted by subtracting the export volume from the import volume. It appears that developed countries generally have a more stable virtual water balance than the developing countries. Countries that are relatively close to each other in terms of geography and development level can have a rather different virtual water balance. Germany, the Netherlands and the UK are net importers whereas France is a net exporter. The US and Canada are net exporter whereas Mexico is a net importer. Although the US has more than three times as much gross virtual water export as Australia, Australia is the country with the largest *net* export of virtual water in the world.

Virtual water flows between world regions

Gross virtual water flows between and within thirteen world regions are presented in Table 3.4. The regions with the largest virtual water export are North and South America. The largest importers are Western Europe and Central and South Asia. The single most important intercontinental water dependency is Central and South Asia (including China and India) annually importing 80 Gm³ of virtual water from North America. This is equivalent to one seventh of the annual runoff of the Mississippi. Ironically, the African continent, not known because of its water abundance, is a net exporter of water to the other continents, particularly to Europe. This can be seen in Figure 3.1, which shows average virtual water balances over the period 1997-2001 at the level of the thirteen world regions. The green regions in the map have net virtual water export, the red ones net virtual water import. The figure shows the biggest virtual water flows between the different regions insofar related to trade in agricultural products.

Table 3.3. Virtual water flows for a few selected countries. Period: 1997-2001.

	Gross virtual water flows ($10^6 m^3/yr$)								Net virtual water import ($10^6 m^3/yr$)			
	Related to the trade of crop products		Related to the trade of livestock products		Related to the trade of industrial products		Total		Related to trade of crop products	Related to trade of livestock products	Related to trade of industrial products	Total
	Export	Import	Export	Import	Export	Import	Export	Import				
Argentina	45952	3100	4178	811	499	1732	50629	5643	-42853	-3367	1233	-44987
Australia	46120	3864	26377	745	501	4399	72998	9007	-42256	-25633	3898	-63991
Bangladesh	771	3670	652	86	162	415	1585	4171	2899	-566	254	2586
Brazil	53713	17467	11911	1907	2211	3694	67835	23068	-36246	-10003	1483	-44767
Canada	48321	16190	17424	4952	29573	14289	95318	35430	-32132	-12472	-15284	-59888
China	17429	36260	5640	15247	49909	11632	72978	63139	18831	9608	-38277	-9839
Egypt	1755	11445	221	1466	729	711	2705	13622	9690	1245	-18	10917
France	43410	40577	13222	11829	21873	19761	78505	72166	-2833	-1393	-2112	-6338
Germany	27630	59751	17432	16062	25416	29757	70478	105570	32121	-1370	4341	35092
India	32411	13941	3406	343	6748	2945	42565	17228	-18470	-3063	-3803	-25337
Indonesia	24750	26917	371	1666	310	1822	25430	30405	2167	1296	1512	4975
Italy	12920	47164	14912	28295	10402	13498	38234	88957	34244	13383	3096	50723
Japan	954	59015	955	20328	4605	18883	6513	98227	58061	19374	14279	91714
Jordan	97	4103	165	462	25	228	287	4794	4006	297	203	4506
Korea Rep.	997	24801	3930	6097	2219	8344	7146	39242	23804	2166	6126	32096
Mexico	11784	26956	5757	13418	3790	9710	21331	50084	15173	7661	5920	28754
Netherlands	34529	48607	15146	7852	7885	12293	57561	68753	14078	-7294	4408	11192
Pakistan	7381	8879	612	98	1526	579	9518	9555	1498	-514	-947	37
Russia	8297	30925	2503	12243	36932	2899	47732	46067	22627	9740	-34032	-1665
South Africa	6326	7752	1312	1019	912	1924	8550	10695	1426	-293	1011	2145
Spain	18252	30483	8541	5972	3753	8520	30545	44975	12231	-2569	4767	14430
Thailand	38429	9761	2856	1761	1655	3596	42940	15117	-28668	-1096	1941	-27823
UK	8773	33742	3786	10163	5113	20321	17672	64226	24968	6378	15208	46554
USA	134623	73129	35484	32919	59195	69763	229303	175811	-61495	-2564	10568	-53491

Table 3.4. Average annual gross virtual water flows between world regions related to the international trade in agricultural products in the period 1997-2001 (Gm³/yr). The grey-shaded cells show the international virtual water flows within a region.

Importer / Exporter	Central Africa	Central America	Central and South Asia	Eastern Europe	Former Soviet Union	Middle East	North Africa	North America	Oceania	South America	South-east Asia	Southern Africa	Western Europe	Total gross export
Central Africa	0.80	0.07	1.73	1.29	0.03	0.26	0.96	0.90	0.06	0.05	1.19	0.17	16.45	23
Central America	0.08	3.13	3.88	0.65	6.14	0.38	0.75	23.98	0.06	0.58	0.23	0.03	10.67	47
Central and South Asia	1.29	0.81	31.53	1.21	4.08	6.67	3.86	4.44	0.37	0.61	16.90	1.37	9.80	51
Eastern Europe	0.01	0.08	0.69	10.77	4.80	2.65	1.08	0.55	0.08	0.10	0.19	0.03	14.15	24
Former Soviet Union	0.01	0.07	3.06	4.47	16.67	5.38	1.26	0.05	0.00	0.30	0.41	0.00	10.54	26
Middle East	0.24	0.11	2.73	0.84	1.46	8.45	3.43	1.01	0.13	0.17	1.86	0.05	6.91	20
North Africa	0.10	0.24	7.09	6.15	2.11	4.32	5.87	8.37	0.17	2.29	3.49	0.52	63.22	98
North America	0.46	40.65	80.18	1.71	2.43	11.22	11.38	35.10	0.96	11.51	13.72	0.79	25.57	201
Oceania	0.34	1.24	29.32	0.33	0.33	6.22	2.13	11.33	12.63	0.67	14.64	1.11	7.76	75
South America	0.39	3.06	19.82	4.23	4.46	8.92	5.08	19.65	0.37	28.09	4.63	1.93	54.44	127
South-east Asia	1.96	0.50	35.57	2.43	1.52	7.75	8.00	10.89	2.49	0.93	26.87	2.54	18.14	93
Southern Africa	1.04	0.06	2.12	0.38	0.19	0.53	0.54	1.12	0.05	0.17	2.41	2.59	7.21	16
Western Europe	1.40	2.60	15.45	18.87	10.56	12.28	14.26	9.79	0.91	2.45	2.61	1.82	183.51	93
Total gross import	7	50	202	43	38	67	53	92	6	20	62	10	245	895

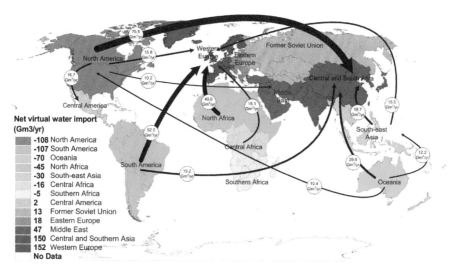

Figure 3.1. Regional virtual water balances and net interregional virtual water flows related to the trade in agricultural products. Period: 1997-2001. Only the biggest net flows (>10 Gm³/yr) are shown.

Figure 3.2. Water scarcity versus water import dependency per country.

Dependence on external water resources

From a water resources point of view one might expect a positive relationship between water scarcity and water import dependency, particularly in the ranges of high water scarcity. Water scarcity is defined here as the country's water footprint – the total water volume needed to produce the goods and services consumed by the people in the country – divided by the country's total renewable water resources (Equation 50). Water import dependency is defined as the ratio of the external water footprint of a country to its total water footprint (Equation 51). The external water footprint of a country refers to the use of water resources in other countries to produce commodities imported into and consumed within the country. Figure 3.2 shows that the relation between water scarcity and water import dependency is not as straightforward as one would expect, although indeed a number of countries – e.g. Kuwait, Qatar, Saudi Arabia, Bahrain, Jordan, Israel, Oman and Lebanon – combine very high water scarcity with very high water import dependency. The water footprints of these countries have largely been externalized.

The reason that the overall picture is more diffuse than one would expect from a water resources point of view, is that under current trade regime water is seldom the dominant factor determining international trade of water-intensive commodities. The relative availability of other input factors – notably land and labour – play a role as well, and also existing national policies, export subsidies and international trade barriers.

Various countries have high water scarcity but low water import dependency. There are different explanatory factors. Yemen, known for overdrawing their limited groundwater resources, for instance has a low water import dependency for the simple reason that they do not have the foreign currency to import water-intensive commodities in order to save domestic water resources. Egypt on the other hand combines high water scarcity and low water import dependency intentionally, aiming at consuming the Nile water to achieve food self-sufficiency.

The water scarcity and use of external water resources for some selected countries are presented in Table 3.5. India has a very high national self-sufficiency ratio (98%), which implies that at present India is only little dependent on the import of virtual water from other countries to meet its national demands. The same is true for the people of China, with a self-sufficiency ratio of 93%. However, India and China have relatively low water footprints per capita (India 980 m^3/cap/yr and China 702 m^3/cap/yr). If the consumption pattern in these countries changes to that like in the US or some Western European countries, they will be facing a severe water scarcity in the future and probably be unable to sustain their high degree of water self-sufficiency.

Table 3.5. Water scarcity and water import dependency for some selected countries. Period: 1997-2001.

Country	Total renewable water resources[1] (Gm³/yr)	Internal water footprint[2] (Gm³/yr)	External water footprint[2] (Gm³/yr)	Total water footprint[2] (Gm³/yr)	Water scarcity (%)	National water self-sufficiency (%)	Water import dependency (%)
Argentina	814	48	3	52	6	94	6
Australia	492	22	5	27	5	82	18
Bangladesh	1,211	112	4	117	10	97	3
Brazil	8,233	216	18	234	3	92	8
Canada	2,902	50	13	63	2	80	20
China	2,897	826	57	883	30	93	7
Egypt	58	56	13	70	119	81	19
France	204	69	41	110	54	63	37
Germany	154	60	67	127	82	47	53
India	1,897	971	16	987	52	98	2
Indonesia	2,838	242	28	270	10	90	10
Italy	191	66	69	135	70	49	51
Japan	430	52	94	146	34	36	64
Jordan	0.9	1.7	4.6	6.3	713	27	73
Korea Rep.	70	21	34	55	79	38	62
Mexico	457	98	42	140	31	70	30
Netherlands	91	4	16	19	21	18	82
Pakistan	223	157	9	166	75	95	5
Russia	4,507	229	42	271	6	84	16
South Africa	50	31	9	40	79	78	22
Spain	112	60	34	94	84	64	36
Thailand	410	123	11	135	33	92	8
UK	147	22	51	73	50	30	70
USA	3,069	566	130	696	23	81	19

[1] FAO (2003)

[2] Chapagain and Hoekstra (2004)

Conclusion

International water dependencies are substantial and are likely to increase with continued global trade liberalization. Today, 16% of global water use is not for producing products for domestic consumption but for making products for export. It means that one sixth of the water problems in the world can be traced back to production for export. Considering this substantial percentage and the upward trend, we suggest that future national and regional water policy studies should include an analysis of international or interregional virtual water flows.

Globalisation of freshwater brings both risks and opportunities. The largest risk is that the indirect effects of consumption are externalized to other countries. While water in agriculture is still priced far below its real cost in most countries, an increasing volume of water is used for processing export products. The costs associated with water use in the exporting country are not included in the price of the products consumed in the importing country. Consumers are generally not aware of and do not pay for the water problems in the overseas countries where their goods are being produced. Efficient and fair trade would require restoring the link between consumers on the one hand and production costs and impacts on the other hand.

Another risk is that national water security of many countries increasingly depends on the import of water-intensive commodities from other countries. Already today, Jordan annually imports a virtual water volume that is five times its own annually renewable water resources. Although saving their own domestic water resources, it increases Jordan's dependency on other nations. Other countries in the same region, such as Kuwait, Qatar, Bahrain, Oman and Israel, but also European countries like the UK, Belgium, the Netherlands, Germany, Switzerland, Denmark, Italy and Malta, have a similar high water import dependency.

An opportunity of reduced trade barriers is that virtual water can be regarded as an alternative source of water. Virtual water import can be used by national governments as a tool to release the pressure on their domestic water resources. In an open world economy, according to international trade theory, the people of a nation will seek profit by trading products that are produced with resources that are abundantly available within the country for products that need resources that are scarcely available. People in countries where water is a comparatively scarce resource, could thus aim at importing products that require a lot of water in their production (water-intensive products) and exporting products or services that require less water (water-extensive products). This import of virtual water (as opposed to import of real water, which is generally too expensive) will relieve the pressure on the nation's own water resources. For water-abundant countries an argumentation can be made for export of virtual water.

Finally, global virtual water trade can physically save water if products are traded from countries with high to countries with low water productivity. For example, Mexico imports wheat, maize and sorghum from the US, which requires 7.1 Gm^3 of water per year in the US. If Mexico would produce the imported crops domestically, it would require 15.6 Gm^3 of water per year. Thus, from a global perspective, the trade of cereals from the US to Mexico saves 8.5 Gm^3/yr. There are also cases where water-intensive commodities flow in the other direction, from countries with low to countries with high water productivity. This is further elaborated in Chapter 4.

Chapter 4

Saving water through global trade[1]

The most direct positive effect of virtual water trade is the water savings it generates in the countries or the regions that import the products. This effect has been widely discussed in virtual water studies since the nineties (Allan, 1999b). These national water savings are equal to the import volumes multiplied by the volumes of water that would have been required to produce the commodities domestically. However, virtual water trade does not only generate water savings for importing countries, it also means water 'losses' for the exporting countries (in the sense that the water cannot be used anymore for other purposes in the exporting countries). The global net effect of virtual water trade between two nations will depend on the actual water volume used in the exporting country in comparison to the water volume that would have been required to produce a commodity in the importing country. There will be net water saving, if the trade is from countries with relatively high water productivity (i.e. commodities have a low virtual water content) to countries with low water productivity (commodities with a high virtual water content). There can be net additional consumption of water if the transfer is from low to high productive sites. The saving can also be realised with transfer of products from low to high productive periods by storage of food, which can be a more efficient and more environmentally friendly way of bridging the dry periods than building large dams for temporary water storage (Renault, 2003).

Virtual water trade between nations is one means of increasing the efficiency of water use in the world. As Hoekstra and Hung (2002; 2005) argue, there are three levels of water use efficiency. At a local level, that of the water user, water use efficiency can be increased by charging prices based on full marginal cost, stimulating water-saving technology, and creating awareness among the water users on the detrimental impacts of water abstractions. At the catchment or river basin level, water use efficiency can be enhanced by re-allocating water to those purposes with the highest marginal benefits. Finally, at the global level, water use efficiency can be increased if nations use their comparative advantage or disadvantage in terms of water availability to encourage or discourage the use of domestic water resources for producing export commodities (respectively stimulate export or import of virtual water). Whereas much research efforts have been dedicated to study water use efficiency at the local and river basin level, little efforts have been done to analyse water use efficiency at global level.

According to the theory of comparative advantage, nations can gain from trade if they concentrate or specialize in the production of goods and services for which they have a comparative advantage, while importing goods and services for which they have a comparative disadvantage (Wichelns, 2001; Wichelns, 2004). The pros and cons of the virtual water trade should be weighed including the opportunity cost of the associated water. Some trade flows may be more beneficial than others purely

---

[1] Based on: Chapagain et al.(2005a, 2005b).

because of the higher opportunity cost of the water being saved. It is relevant for instance to look whether water saved is *blue* or *green* water. Green water is the productive use of rainfall in crop production, which, in general, has a lower opportunity cost compared to the blue water use (i.e. irrigation).

The average global volume of virtual water flows related to the international trade in agricultural products was 1263 Gm^3/yr in the period 1997-2001 (Chapter 3). This estimate is based on the virtual water content of the products in the exporting countries. It would be interesting to see the volume of virtual water traded internationally based on the virtual water content of the products in the importing countries. Zimmer and Renault (2003) estimated this as 1340 Gm^3 /yr related to the international trade in crop and livestock products in the year 2000. These studies only present a partial view of the global or national savings.

An estimate of global virtual water trade and resulting global water saving was done by Oki *et al.* (2003) and Oki and Kanae (2004; 2003). They estimated the global sum of virtual water exports on the basis of the virtual water content of the products in the exporting countries (683 Gm^3/yr) and the global sum of virtual water imports on the basis of the virtual water content of the products in the importing countries (1138 Gm^3/yr). This saves 455 Gm^3/yr as a result of food trade. Their study is severely limited with respect to the methodology followed in calculating the virtual water content of a product. First, they have assumed a constant global average crop water requirement throughout the world, being 15 mm/day for rice and 4 mm/day for maize, wheat and barley. Thus the climatic factor, which plays a major role in the crop water requirement of a crop, is completely neglected. Secondly, they did not take into account the role of the crop coefficient, which is the major limiting factor determining the evaporation from a crop at different stages of crop growth. The global virtual water flows and the resulting water savings as calculated in these studies are limited to the international trade of four major crops (maize, wheat, rice and barley) only.

In this chapter the global and national water savings have been quantified and analysed with proper accounting of climate, yield, and cropping pattern per crop per country for the period 1997-2001. The study covers the international trade of all major crop and livestock products.

The virtual water content of a product is calculated using the methodology as developed by Hoekstra and Hung (2002; 2005) and Chapagain and Hoekstra (2003a; 2003b). First the virtual water content (m^3/ton) of the primary crop is calculated based on crop water requirement and yield in the producing country. The crop water requirement is calculated using the methodology developed by FAO (Allen *et al.*, 1998). The calculation is done using the climate data of the producing country and the specific cropping pattern of each crop per country. The virtual water content (m^3/ton) of live animals has been calculated based on the virtual water content of their feed and the volumes of drinking and service water consumed during their lifetime. The virtual water content of processed products is calculated based on product fractions (ton of crop product obtained per ton of primary crop or live animal) and value fractions (the market value of one crop or livestock product divided by the aggregated market value of all products derived from one primary crop or live animal). The product fractions have been taken from the commodity trees in FAO (2003f). The value fractions have been calculated based on the market prices of the various products. The global average market prices of the different products for the period 1997-2001 have been calculated using trade data from the International Trade Centre (ITC, 2004).

The national water saving ΔS_n (m^3/yr) of a country as a result of trade of product p is the calculated by multiplying the net import of product (ton/yr) by its virtual water content at the importing country (Equation 53). Obviously, ΔS_n can have a negative sign, which means a net water loss instead of a saving.

The global water saving ΔS_g (m^3/yr) through the trade of a product p from an exporting country to an importing country is calculated by multiplying the volume of product traded with the difference in virtual water content in the trading countries (Equation 54). If the export of product is from sites where its virtual water content is higher to the sites where it has lower virtual water content there is a net loss fromm the global system instead of the saving. The total global water saving can be obtained by summing up the global savings of all trades ΔS_g. By definition, the total global water saving is also equal to the sum of the national savings of all countries ΔS_n.

The case of global water saving is illustrated with an example of the import of husked rice in Mexico from the USA in Figure 4.1.

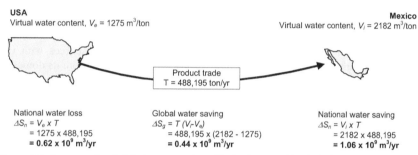

Figure 4.1. An example of global water saving with the import of husked rice in Mexico from USA.

The case of global water loss is shown with an example of export of broken rice from Thailand to Indonesia in Figure 4.2. For the computation of the total water saving that is made by international trade of agricultural products, the calculation has been carried out for 285 crop products and 123 livestock products as reported in the database PC-TAS (ITC, 2004) which covers international trade between 243 countries for 1997-2001.

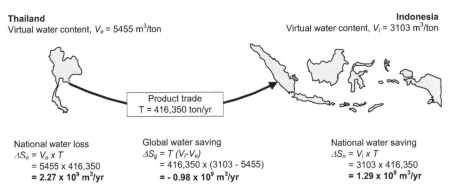

Thailand
Virtual water content, V_e = 5455 m³/ton

Indonesia
Virtual water content, V_i = 3103 m³/ton

Product trade
T = 416,350 ton/yr

National water loss
$\Delta S_n = V_e \times T$
= 5455 x 416,350
= **2.27 x 10⁹ m³/yr**

Global water saving
$\Delta S_g = T (V_i - V_e)$
= 416,350 x (3103 - 5455)
= **- 0.98 x 10⁹ m³/yr**

National water saving
$\Delta S_n = V_i \times T$
= 3103 x 416,350
= **1.29 x 10⁹ m³/yr**

Figure 4.2. An example of global water loss with the import of broken rice in Indonesia from Thailand.

National water savings

A large number of countries are saving their national water resources with the international trade of agricultural products. Japan saves 94 Gm³/yr from its domestic water resources, Mexico 65 Gm³/yr, Italy 59 Gm³/yr, China 56 Gm³/yr and Algeria 45 Gm³/yr. The global picture of national savings is presented in Figure 4.3. The water savings shown in the figure are net water savings. A net national water saving is the result of a gross water saving and a gross water loss. The driving forces behind international trade of water-intensive products can be water scarcity in the importing countries, but often other factors such as scarcity of fertile land or other resources play a decisive role (Yang *et al.*, 2003). As a result, realised national water savings can only partially be explained through national water scarcity.

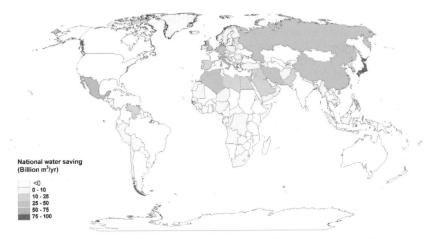

National water saving
(Billion m³/yr)
<0
0 - 10
10 - 25
25 - 50
50 - 75
75 - 100

Figure 4.3. National water savings related to international trade of agricultural products. Period 1997-2001.

The national water saving has different implications per country. Though Germany saves 34 Gm^3/yr, it may be less important from a national policy making perspective because the major products behind the saving are stimulant crops (tea, coffee and cocoa) which Germany would otherwise not produce itself. If the import of stimulants is reduced, it may not create any additional pressure on the water resources in Germany. However, for Morocco, where import of cereal crop products is the largest national water saver, shifting from import to domestic production would create an additional pressure of 21 Gm^3/yr on its national water resources. The nations that save most water through international trade of agricultural products and the main products behind the savings are presented in Table 4.1.

Table 4.1. Nations with the largest net water saving as a result of international trade of agricultural products. Period 1997-2001.

Countries	Net national water saving (Gm^3/yr)	Major partners (Gm^3/yr)	Major product categories (Gm^3/yr)
Japan	94	USA (48.9), Australia (9.6), Canada (5.4), Brazil (3.8), China (2.6)	Cereal crops (38.7), oil-bearing crops (23.2), livestock (16.1), stimulants (9.2)
Mexico	65	USA (54.0), Canada (5.1)	Livestock (31.0), oil-bearing crops (20.5), cereal crops (19.3)
Italy	59	France (14.6), Germany (6.0), Brazil (5.4), Netherlands (4.4), Argentina (3.1), Spain (3.1)	Livestock (23.2), cereal crops (15.2), oil-bearing crops (12.9), stimulant (8.1)
China	56	USA (17.4), Brazil (8.3), Argentina (8.3), Canada (3.6), Italy (3.4), Australia (3.2), Thailand (2.6)	Livestock (27.5), oil-bearing crops (32.6)
Algeria	45	Canada (10.8), USA (7.6), France (7.1), Germany (4.0), Argentina (1.6)	Cereal crops (33.7), oil-bearing crops (4.0), livestock (3.4)
Russian Fed.	41	Kazakhstan (5.2), Germany (4.4), USA (4.1), Ukraine (3.4), Brazil (3.3), Cuba (2.4), France (1.9), Netherlands (1.9)	Livestock (15.2), cereal crops (7.1), sugar (6.9), oil-bearing crops (4.3), stimulant (3.8), fruits (2.3)
Iran	37	Brazil (8.3), Argentina (8.1), Canada (7.7), Australia (6.0), Thailand (2.2), France (2.0)	Cereal crops (22.5), oil-bearing crops (15.1), sugar (1.6)
Germany	34	Brazil (8.3), Cote d'Ivoire (5.3), Netherlands (5.0), USA (4.2), Indonesia (3.3), Argentina (2.2), Colombia (2.1)	Stimulants (21.8), oil-bearing crops (15.0), fruits(3.4), nuts (2.3)
Korea Rep.	34	USA (15.6), Australia (3.6), Brazil (2.2), China (1.5), India (1.4), Malaysia (1.2), Argentina (1.1)	Oil-bearing crops (14.3), cereal crops (12.8), livestock (2.3), sugar (1.9), stimulants (1.5)
UK	33	Netherlands (5.3), France (3.7), Brazil (2.8), Ghana (1.9), USA (1.8), Cote d'Ivoire (1.5), Argentina (1.4)	Oil-bearing crops (10.1), stimulants (9.5), livestock (5.2)
Morocco	27	USA (7.8), France (6.4), Argentina (3.3), Canada (2.2), Brazil (1.2), Turkey (0.8), UK (0.8)	Cereal crops (20.9), oil-bearing crops (4.4)

For an importing country it is not relevant whether products are consuming green or blue water in the exporting country. The importing country is more interested to see what volume and kind of water is being saved from its own resources by the import. And it is further important to see whether the water thus saved has higher marginal benefits than the additional cost to import these products.

As an example, Figure 4.4 shows the national water saving of Egypt as a result of the import of wheat. In Egypt, the mean rainfall is only 18 mm/yr. Almost all agriculture in Egypt is irrigated. At present, Egypt and Sudan base their water resources plan on the agreed division of water by the 1959 Nile water agreement between Sudan and Egypt. However, future developments in upstream countries will have to be taken into account. Disputes over the distribution of water of the Nile could become a potential source of conflict and contention. The expansion of irrigation in the basin will require basin-wide cooperation in the management of water resources to meet increasing demands and to face the associated environmental consequences.

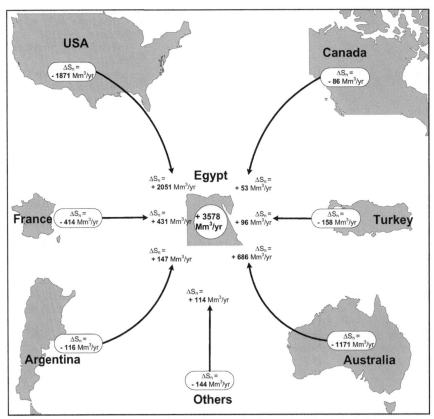

Figure 4.4. National water saving related to the net wheat import of Egypt. Period 1997-2001.

In this context, the import of wheat in Egypt is contributing to national water saving of 3.6 Gm^3/yr which is about seven percent of the total volume of water Egypt is entitled to according to the 1959 agreement. The national saving is made with the investment of foreign exchange of 593 million US$/yr (ITC, 2004). Hence, from an economic point of view, the opportunity cost of the resources being saved (such as land, water and labour) should be more or at least equal to the price paid for it. If the opportunity cost of land and labour approaches zero, the opportunity cost of water being saved should be more than 0.17 US$/m^3. But the import of wheat in Egypt should be assessed including other factors of production such as land and labour. In Egypt fertile land is also a major scarce resource. The pressure to increase the land area with reclamation is released to some extent by the wheat import but on the other hand the import is made at the cost of employment lost. Greenaway *et al.* (1994) and Wichelns (2001) have shown that the production of wheat has a comparative disadvantage in Egypt. As the saving is completely in blue water, the marginal utility of the saved water may justify the import economically.

National water losses

Whereas import of agricultural products implies that national water resources are saved, export of agricultural products entails that national water resources are lost. The term 'national water loss' is used in this paper to refer to the fact that water used for producing commodities that are consumed by people in other countries is not available anymore for in-country purposes. The term 'water loss' is used here as the opposite of 'water saving'. The terms 'loss' and 'saving' are not to be interpreted in terms of economic loss or saving, but in a physical manner (Equation 53). Calculated national water savings and losses become valued positive or negative in an economic sense depending on the context. Water losses as defined here are negative in economic sense only if the benefit in terms of foreign earning does not outweigh the costs in terms of opportunity cost and negative externalities left at the site of production.

The nations with the largest net water loss are the USA (92 Gm^3/yr), Australia (57 Gm^3/yr), Argentina (47 Gm^3/yr), Canada (43 Gm^3/yr), Brazil (36 Gm^3/yr) and Thailand (26 Gm^3/yr). Figure 4.5 shows the water losses of all countries that have a net water loss due to the production for export.

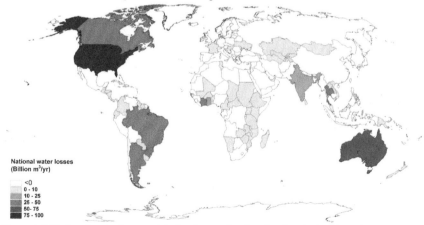

National water losses
(Billion m^3/yr)

<0
0 - 10
10 - 25
25 - 50
50 - 75
75 - 100

Figure 4.5. National water losses related to international trade of agricultural products. Period 1997-2001.

The list of nations with the largest net water loss through the international trade of agricultural products is presented in Table 4.2. The main products behind the national water loss from the USA are oil-bearing crops and cereal crops. These products are partly produced rain-fed and partly irrigated. However the loss from Cote d'Ivoire and Ghana is mainly from the export of stimulants, which are almost entirely rain-fed. The use of green water has no major competition with other uses in these countries. This type of loss to the national water resources is unlikely to be questionable from an economic perspective, because the opportunity costs of this water are low. The concern is limited to the environmental impacts, which are generally not included in the price of the export products.

Table 4.2. Nations with the largest net water loss as a result of international trade of agricultural products. Period 1997-2001.

Countries	Net national water loss (Gm^3/yr)	Major partners (Gm^3/yr)	Major product categories (Gm^3/yr)
USA	92	Japan (29.2), Mexico (26.8), China (14.1), Korea Rep (10.1), Taiwan (8.4), Egypt (3.8), Spain (3.7)	Oil-bearing crops (65.2), cereal crops (45.4), livestock (7.8)
Australia	57	Japan (13.7), China (6.0), USA (5.7), Indonesia (4.7), Korea Rep (3.9), Iran (3.3)	Cereal crops (23.1), livestock (24.3), oil-bearing crops (6.8), sugar (4.3)
Argentina	47	Brazil (6.7), China (3.7), Spain (2.4), Netherlands (2.2), Italy (2.1), USA (2.0), Iran (1.9)	Oil-bearing crops (29.9), cereal crops (12.8), livestock (3.7)
Canada	43	USA (12.4), Japan (7.9), China (5.2), Iran (3.7), Mexico (3.4), Algeria (2.1)	Cereal crops (29.3), livestock (12.3), oil-bearing crops (9.6)
Brazil	36	Germany (5.8), USA (5.3), China (4.5), Italy (4.2), France (4.2), Netherlands (3.9), Russian Fed (2.8)	Oil-bearing crops (17.7), stimulants (15.8), sugar (9.0), livestock (9.3)
Cote d'Ivoire	32	Netherlands (5.7), France (4.7), USA (4.5), Germany (4.1), Italy (1.7), Spain (1.5), Algeria (1.4)	Stimulants (32.9), oil-bearing crops (1.5)
Thailand	26	Indonesia (4.7), China (4.4), Iran (2.6), Malaysia (2.5), Japan (2.3), Senegal (1.8), Nigeria (1.7)	Cereal crops (23.6), Sugar (5.1), roots and tuber (2.5)
Ghana	17	Netherlands (3.6), UK (3.3), Germany (1.7), Japan (1.6), USA (1.3), France (1.0)	Stimulants (19.1)
India	13	China (2.4), Saudi Arabia (2.0), Korea Rep (1.8), Japan (1.6), Russian Fed (1.3), France (1.3), USA (1.3)	Cereal crops (6.1), stimulants (3.2), livestock (3.0), oil-bearing crops (1.8)
France	9	Italy (6.4), Belgium-Luxembourg (3.8), UK (2.8), Germany (2.1), Greece (1.6), Algeria (1.4), Morocco (1.1)	Cereal crops (21.9), sugar (4.6), livestock (4.2)
Vietnam	8	Indonesia (2.3), Philippines (1.7), Ghana (0.4), USA (0.4), Germany (0.4), Senegal (0.4), Singapore (0.4)	Cereal crops (6.8), stimulants (2.7)

The national water losses from France, Vietnam and Thailand are mainly the result of cereal crop products. Particularly the example of rice export from Thailand is interesting from blue water and opportunity cost perspective (Figure 4.6). Thailand exports 27.8 Gm^3/yr of water in the form of rice. The monetary equivalent of rice export is 1556 million US\$/yr (ITC, 2004). Hence, from the loss of its national water, Thailand is generating foreign exchange of 0.06 US\$/m^3. The water loss is partly from blue water resources and partly from green water resources. As rice cultivation in Thailand is done during the rainy season, the share of green water is quite considerable in the virtual water content of the rice. Here, one needs to remember that the benefits of rice export should be attributed to all the resources consumed in the production process such as water, land and labour. If the contribution of rainfall is 50% to the total evaporative demand of the crop, and if other resources have zero opportunity cost (which is not the case) the opportunity cost of rice export from Thailand approaches 0.12 US\$/m^3 of blue water. Though it is a crude estimation of opportunity cost of rice export, it indicates that the volume of national water loss could have produced higher economic benefits to the nation.

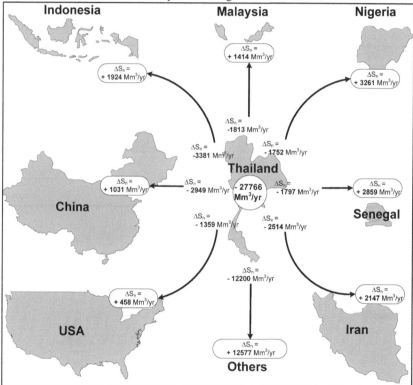

Figure 4.6. National water loss related to the net rice export of Thailand. Period 1997-2001.

Global water savings

Considering the international trade flows between all major countries of the world and looking at the major agricultural products being traded (285 crop products and 123 livestock products), it has been calculated that the global water saving by trade in agricultural products is 352 Gm^3/yr (Table 4.3). This volume equals 28% of the international virtual water flows related to agricultural product trade and 6% of the global volume of water used for agricultural production (which is 6391 Gm^3/yr, Chapter 3). The trade flows that save more than 0.5 Gm^3/yr are shown in Figure 4.7. The trade flows between USA-Japan and USA-Mexico are the biggest global water savers. The contribution of different product groups to the total global water saving is presented in Figure 4.8. Cereal crop products form the largest group responsible for the total global water saving, with a saving of 222 Gm^3/yr, followed by oil-bearing crops (68 Gm^3/yr, mainly soybeans) and livestock products (45 Gm^3/yr). The cereal group is composed of wheat (103 Gm^3/yr), maize (68 Gm^3/yr), rice (21 Gm^3/yr), barley (21 Gm^3/yr), and others (9 Gm^3/yr).

Table 4.3. Global virtual water flows and water savings. Period 1997-2001.

	Related to trade of agricultural products		
	Crop products (Gm^3/yr)	Livestock products (Gm^3/yr)	Total (Gm^3/yr)
Global sum of virtual water exports, assessed on the basis of the virtual water content of the products in the exporting countries (Gm$^{3/}$yr)	979	275	1254
Global sum of virtual water imports, assessed on the basis of the virtual water content of products if produced in the importing countries (Gm$^{3/}$yr)	1286	320	1646
Global water saving (Gm$^{3/}$yr)	307	45	352
Saving compared to the sum of international virtual water flows (%)	34%	16%	30%
Saving compared to the global water use for agricultural products (%)	5.3%	0.7%	6%

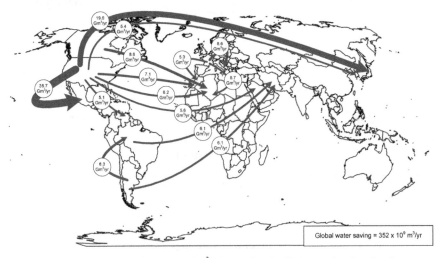

Figure 4.7. Global water savings (>5.0 Gm^3/yr) associated with international trade of agricultural products. Period 1997-2001.

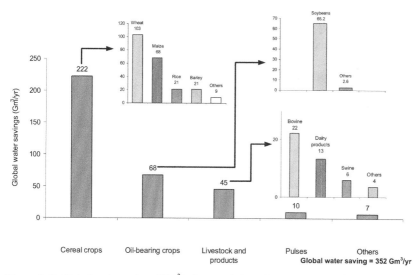

Figure 4. 8. Global water savings (Gm^3/yr) per traded product category. Period 1997-2001.

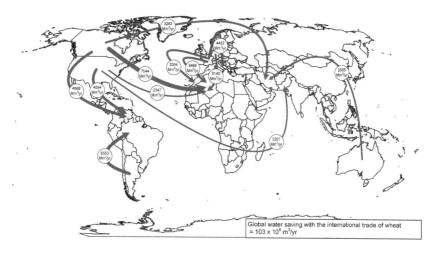

Figure 4. 9. Global water savings (>2.0 Gm3/yr) associated with the international trade of wheat. Period 1997-2001.

The largest global water savings by wheat trade are occurring as a result of wheat import in the Middle East and North African from Western Europe and North America. Figure 4.9 shows the wheat trade flows saving more than 2 Gm^3 of water per year. Maize imports in Japan alone are responsible for 15 Gm^3/yr of global water saving. The global saving of water as a result of maize trade is mainly from the export of maize from USA. Figure 4.10 shows the maize trade flows saving more than 1 Gm^3/yr. Figure 4.11 shows the global water savings above 0.5 Gm^3/yr as a result of rice trade. As the production is more favourable (climate and culture) in

South-east Asia, the largest savings are from the export from this region to the Middle East and West Africa. The major saving through the trade of rice is between Thailand-Iraq, Thailand-Nigeria, Syria-Nigeria, and China-Indonesia.

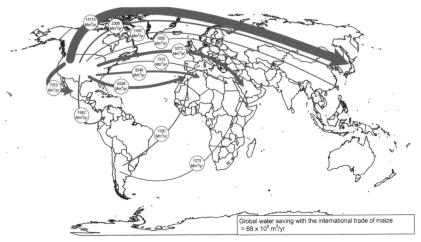

Figure 4.10. Global water savings (>1.0 Gm³/yr) associated with the international trade of maize. Period 1997-2001.

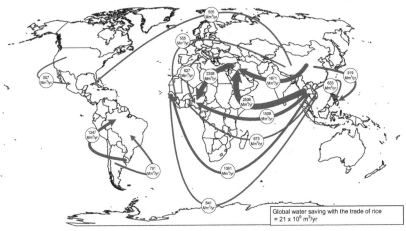

Figure 4.11. Global water savings (>0.5 Gm³/yr) associated with the international trade of rice. Period 1997-2001.

Considering the import of wheat in Egypt, one can see that this contributes to global water saving in some cases and global water loss in other cases (Figure 4.4). The import from USA, France and Argentina is globally saving water by 0.23 Gm³/yr, whereas the import of wheat from Canada, Turkey and Australia results in a global water loss of 0.58 Gm³/yr. Though Egypt's import of wheat saves national water resources by 3.6 Gm³/yr, it results in a net global water loss of 0.4 Gm³/yr. The crop water requirement in Egypt is relatively high compared to its trading partners, but this is partially compensated by a relatively high wheat yield, which is more than twice the global average (Table 4.4). As a result, water productivity (water use per unit of product) in wheat production in Egypt is higher than in Canada, Turkey and Australia. However, wheat production in Egypt is using scarce blue water resources and the partner countries are making use of the effective rainfall (green water). The net global water loss related to the wheat export from Canada etc. to Egypt results from the fact that the volume of *blue* water resources that would have been required in Egypt to produce domestically is smaller than the volume of *green* water resources actually used in Canada etc. Blue and green water resources fundamentally differ in terms of possible application and thus opportunity cost. For further analysis and interpretation of figures on global water savings or losses it is thus important to split up these figures into a blue and green water component.

Table 4.4. Crop water requirements, crop yields and the virtual water content of wheat in Egypt and its major trade partners. Period 1997-2001.

	Crop water requirement (mm/crop period)	Wheat yield (ton/ha)	Virtual water content (m^3/ton)
Argentina	179	2.4	738
Australia	309	1.9	1588
Canada	339	2.3	1491
Egypt	570	6.1	930
France	630	7.0	895
Turkey	319	2.1	1531
USA	237	2.8	849
Global average		2.7	1334

A second example elaborated here is the trade of maize from the USA to Japan. The global water saving from this trade is 15.4 Gm^3/yr. The evaporative demand of maize in Japan (367 mm/crop period) is comparable with that in the USA (411 mm/crop period), but the crop yield in the USA (8.4 ton/ha) is significantly higher than in Japan (2.5 ton/ha), so that the virtual water content of maize in Japan is 3 times higher than in the USA. Saving domestic water resources is not the only positive factor for Japan. If Japan would like to grow the quantity of maize which is now imported from the USA, it would require 6 million hectare of additional cropland. This is a lot given the scarcity of land in Japan.

A third case considered here is rice export from Thailand. Though Thailand looses water by exporting to Nigeria and Senegal by 1.7 Gm^3/yr and 1.8 Gm^3/yr respectively, it is saving water globally as the national water savings in Nigeria (3.2 Gm^3/yr) and Senegal (2.9 Gm^3/yr) are higher than the losses in Thailand (Figure 4.6). The main reason behind the global saving related to the trade between Thailand and Nigeria, is that rice yield in Thailand is 1.7 times higher than in Nigeria (Table 4.5). These two countries have crop water requirements of comparable magnitude (1000 mm/crop period). On the contrary, the main reason behind the global water saving by the trade between Thailand and Senegal, which both have a crop yield in the order of 2.5 ton/ha, is the difference in the crop water requirements in Thailand (945 mm/crop period) and Senegal (1523 mm/crop period). The export of rice from Thailand to five other trading partners (China, Indonesia, Iran, Malaysia and USA) is creating a global water loss of 5 Gm^3/yr. National water loss in Thailand is greater than the corresponding national water savings in these countries. This is due to the fact that rice yield in Thailand is low if compared to the countries where it exports to.

Table 4.5. Crop water requirements, crop yields and the virtual water content of rice in Thailand and its major trade partners. Period 1997-2001.

	Crop water requirement (mm/crop period)	Rice yield (ton/ha)	Virtual water content (m³/ton)
China	830	6.3	1321
Indonesia	932	4.3	2150
Iran	1306	4.1	3227
Malaysia	890	3.0	2948
Nigeria	1047	1.5	7036
Senegal	1523	2.5	6021
Thailand	945	2.5	3780
USA	863	6.8	1275
Global average		3.9	2291

Global blue water savings at the cost of green water losses

The global water saving ΔS_g is made up of a *global blue water saving* ($\Delta S_{g,b}$) and a *global green water saving* ($\Delta S_{g,g}$) component:

$$\Delta S_g = T \times (V_i - V_e)$$
$$= T \times \left((V_{g,i} + V_{b,i}) - (V_{g,e} + V_{b,e}) \right) \tag{55}$$
$$= T \times (V_{g,i} - V_{g,e}) + T \times (V_{b,i} - V_{b,e})$$
$$= \Delta S_{g,g} + \Delta S_{g,b}$$

Even if there is a net global water loss from a trade relation, there might be a saving of blue water at the cost of a greater loss of green water or vice versa. The case is elaborated with the example of Egypt's wheat trade. The virtual water content of wheat in Egypt is 930 m³/ton. This is all blue water; the green component of the virtual water content of wheat is zero. Suppose that Egypt is importing T ton/yr of wheat from Australia. The virtual water content of wheat in Australia is 1588 m³/ton. Wheat production in Australia is not 100% irrigated; it is assumed here that a fraction f of the virtual water content of wheat in Australia is green water. There is net global loss of $658T$ m³/yr in this trade.

$$\Delta S_g = T \times (V_i - V_e)$$
$$= T \times (930 - 1588)$$
$$= -658T$$

The global *green* water saving, $\Delta S_{g,g}$ (m³/yr), in this case is always negative:

$$\Delta S_{g,g} = T \times (V_{g,i} - V_{g,e})$$
$$= T \times (0 - f \times 1588)$$
$$= -1588T$$

However, whether the global *blue* water saving $\Delta S_{g,b}$ (m³/yr) is positive or negative depends upon the fraction f in the exporting country:

$$\Delta S_{g,b} = T \times \left(V_{b,i} - V_{b,e} \right)$$
$$= T \times \left(930 - (1 - f)1588 \right)$$
$$= T \times \left(-658 + 1588 f \right)$$

There is net gain in global blue water resources as long as the blue water component of Australian wheat is smaller than in Egypt, i.e. if the fraction f in Australia is larger than 0.42. In a case of extreme drought, if the effective rainfall in Australia for wheat is zero ($f=0$) and all the evaporative demand is met by irrigation, all the losses are in blue water resources, which is $658T$ m^3/yr. In another extreme example, when the full evaporative demand of wheat in Australia is met by effective rainfall, so that no irrigation water is used ($f=1$), the global loss of green water will be $1588T$, but we obtain a net global gain of blue water of $930T$ m^3/yr. Here, the gain in blue water is obtained at the cost of green water.

Since blue water resources are generally scarcer than green water resources, global water losses can be positively evaluated if still blue water resources are being saved. The classical example of trade that makes sense from both water resources and economic point of view is when predominantly rain-fed crop or livestock products from humid areas are imported into a country where effective rainfall is negligible. Also the import of products that originate from semi-arid countries that apply supplementary irrigation can be beneficial from a global point of view, because supplementary irrigation increases yields often more than double, a profitable situation that can never be achieved in arid countries where effective rainfall is too low to allow for supplementary irrigation, so that full irrigation is the only option.

Conclusion

The volume of global water saving from the international trade of agricultural products is 352 Gm^3/yr (average over the period 1997-2001). The largest savings are from international trade of crop products, mainly cereals (222 Gm^3/yr) and oil crops (68 Gm^3/yr), owing to the large regional differences in virtual water content of these products and the fact that these products are generally traded from water efficient to less water efficient regions. Since there is smaller variation in the virtual water content of livestock products, the savings by trade of livestock products are less.

The export of a product from a water efficient region (relatively low virtual water content of the product) to a water inefficient region (relatively high virtual water content of the product) saves water globally. This is the physical point of view. Whether trade of products from water efficient to water inefficient countries is beneficial from an economic point of view, depends on a few additional factors, such as the character of the water saving (blue or green water saving), and the differences in productivity with respect to other relevant input factors such as land and labour. Besides, international trade theory tells that it is not the absolute advantage of a country that indicates what commodities to produce but the relative advantage (Wichelns, 2004). The decision to produce locally or to import from other sites should be made on the basis of the marginal value or the utility of the water being saved at the consumption site compared to the cost of import.

Saving domestic water resources in countries that have relative water scarcity by the mechanism of virtual water import (import of water-intensive products) looks very attractive. There are however a number of drawbacks that have to be taken into account as well. Saving domestic water through import should explicitly be seen in the context of:

- the need to generate sufficient foreign exchange to import food which otherwise would be produced domestically;
- the risk of moving away from food self sufficiency that associates with the fear of being held to political ransom;
- increased urbanization in importing countries as import reduces employment in the agricultural sector;
- reduced access of the poor to food; and
- increased risk of environmental impact in exporting countries, which is generally not accounted for in the price of the imported products.

Enhanced virtual water trade to optimise the use of global water resources can relieve the pressure on water scarce countries but may create additional pressure on the countries that produce the water-intensive commodities for export. The potential water saving from global trade is only sustainable if the prices of the export commodities truly reflect the opportunity costs and negative environmental externalities in the exporting countries. Otherwise the importing countries simply gain from the fact that they would have to bear the cost of water depletion if they would produce domestically whereas the costs remain external if they import the water-intensive commodities instead.

Since an estimated 16% of the global water use is not for domestic consumption but for export, global water use efficiency becomes an important issue with increasing globalisation of trade. Though international trade is seldom done to enhance global water productivity, there is an urgent need to address the increasing global water scarcity problem.

Chapter 5

Water footprints of nations: water use by people as a function of their consumption pattern[1]

Databases on water use traditionally show three columns of water use: water withdrawals in the domestic, agricultural and industrial sector respectively (FAO, 2003a; Gleick, 1993; Shiklomanov, 2000). A water expert being asked to assess the water demand in a particular country will generally add the water withdrawals for the different sectors of the economy. Although useful information, this does not tell much about the water actually needed by the people in the country in relation to their consumption pattern. The fact is that many goods consumed by the inhabitants of a country are produced in other countries, which means that it can happen that the real water demand of a population is much higher than the national water withdrawals do suggest. The reverse can be the case as well: national water withdrawals are substantial, but a large amount of the products are being exported for consumption elsewhere.

In 2002, the water footprint concept was introduced in order to have a consumption-based indicator of water use that could provide useful information in addition to the traditional production-sector-based indicators of water use (Hoekstra and Hung, 2002). The water footprint of a nation is defined as the total volume of freshwater that is used to produce the goods and services consumed by the people of the nation. Since not all goods consumed in one particular country are produced in that country, the water footprint consists of two parts: use of domestic water resources and use of water outside the borders of the country.

The water footprint has been developed in analogy to the ecological footprint concept as was introduced in the 1990s (Rees, 1992; Wackernagel et al., 1997; Wackernagel and Rees, 1996). The 'ecological footprint' of a population represents the area of productive land and aquatic ecosystems required to produce the resources used, and to assimilate the wastes produced, by a certain population at a specified material standard of living, wherever on earth that land may be located. Whereas the 'ecological footprint' thus quantifies the *area* needed to sustain people's living, the 'water footprint' indicates the *water* required to sustain a population.

The water footprint concept is closely linked to the virtual water concept. Virtual water is defined as the volume of water required to produce a commodity or service. The concept was introduced by Allan in the early 1990s (Allan, 1993; 1994) when studying the option of importing virtual water (as opposed to real water) as a partial solution to problems of water scarcity in the Middle East. Allan elaborated on the idea of using virtual water import (coming along with food imports) as a tool to release the pressure on the scarcely available domestic water resources. Virtual water import thus becomes an alternative water source, next to endogenous water

---

[1] Based on: Chapagain and Hoekstra (2004); (Hoekstra and Chapagain, 2005a); Hoekstra and Chapagain (2005c) accepted for publication in Water Resources Management.

sources. Imported virtual water has therefore also been called 'exogenous water' (Haddadin, 2003).

When assessing the water footprint of a nation, it is essential to quantify the flows of virtual water leaving and entering the country. If one takes the use of domestic water resources as a starting point for the assessment of a nation's water footprint, one should subtract the virtual water flows that leave the country and add the virtual water flows that enter the country.

In this Chapter the water footprint of nations is assessed and analysed for the period of 1997-2001. The study builds on two earlier studies. Hoekstra and Hung (2002; 2005) have quantified the virtual water flows related to the international trade of crop products. Chapagain and Hoekstra (2003a) have done a similar study for livestock and livestock products. The concerned time period in these two studies is 1995-99. The present study refines the earlier studies by making a number of improvements and extensions.

A nation's water footprint has two components, the internal and the external water footprint (Equation 45). The internal water footprint is the use of domestic water resources to produce goods and services consumed by inhabitants of the country. It is the sum of the total water volume used from the domestic water resources in the national economy *minus* the volume of virtual water export to other countries insofar related to export of domestically produced products (Equation 46). The agricultural water use is taken equal to the evaporative water demand of the crops. The agricultural water use includes both effective rainfall (the portion of the total precipitation which is retained by the soil and used for crop production) and the part of irrigation water used effectively for crop production. Here irrigation losses are not included in the term of agricultural water use assuming that they largely return to the resource base and thus can be reused.

The external water footprint of a country is the annual volume of water resources used in other countries to produce goods and services consumed by the inhabitants of the country concerned. It is equal to the so-called virtual water import into the country *minus* the volume of virtual water exported to other countries as a result of re-export of imported products (Equation 47). Both the internal and the external water footprint include the use of *blue water* (ground and surface water) and the use of *green water* (moisture stored in soil strata).

The use of domestic water resources comprises water use in the agricultural, industrial and domestic sectors. For the latter two sectors we have used data from AQUASTAT (FAO, 2003a). Though significant fractions of domestic and industrial water withdrawals do not evaporate but return to either the groundwater or surface water system, these return flows are generally polluted, so that they have been included in the water footprint calculations. The total volume of water use in the agricultural sector has been calculated in this study based on the total volume of crop produced and its corresponding virtual water content. For the calculation of the virtual water content of crop and livestock products we have used the methodology as described in Chapagain and Hoekstra (2004). In summary, the virtual water content (m^3/ton) of primary crops has been calculated based on crop water requirements and yields. Crop water requirement have been calculated per crop and per country using the methodology developed by FAO (Allen *et al.*, 1998). The virtual water content of crop products is calculated based on product fractions (ton of crop product obtained per ton of primary crop) and value fractions (the market value of one crop product divided by the aggregated market value of all crop products derived from one primary crop). The virtual water content (m^3/ton) of live animals has been calculated based on the virtual water content of their feed and the volumes of drinking and service water consumed during their lifetime. The virtual water content for eight major animal categories has been calculated: beef cattle, dairy cows, swine, sheep, goats, fowls/poultry (meat purpose), laying hens and horses. The calculation of the virtual water content of livestock products is again based on product fractions and value fractions.

Virtual water flows between nations have been calculated by multiplying commodity trade flows by their associated virtual water content (Equation 40). We have taken into account the trade between 243 countries for which international trade data are available in the Personal Computer Trade Analysis System of the International Trade Centre, produced in collaboration with UNCTAD/WTO. It covers trade data from 146 reporting countries disaggregated by product and partner countries (ITC, 2004). We have carried out calculations for 285 crop products and 123 livestock products. The virtual water content of an industrial product can be calculated in a similar way as described earlier for agricultural products. There are however numerous categories of industrial products with a diverse range of production methods and detailed standardised national statistics related to the production and consumption of industrial products are hard to find. As the global volume of water used in the industrial sector is only 716 Gm^3/yr ($\approx$ 10% of total global water use, Chapter 3), we have – per country – simply calculated an average virtual water content per dollar added value in the industrial sector (m^3/US$) as the ratio of the industrial water withdrawal (m^3/yr) in a country to the total added value of the industrial sector (US$/yr), which is a component of the Gross Domestic Product (Equation 38 and 39).

Water needs by product

The total volume of water used globally for crop production is 6390 Gm^3/yr at field level. Rice has the largest share in the total volume water used for global crop production. It consumes about 1359 Gm^3/yr, which is about 21% of the total volume of water used for crop production at field level. The second largest water consumer is wheat (12%). The contribution of some major crops to the global water footprint insofar related to food consumption is presented in Figure 5.1.

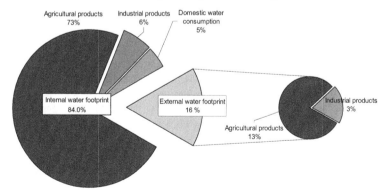

Figure 5.1. Contribution of different crops to the global water footprint.

Although the total volume of the world rice production is about equal to the wheat production, rice consumes much more water per ton of production. The difference is due to the higher evaporative demand for rice production. As a result, the global average virtual water content of rice (paddy) is 2291 m³/ton and for wheat 1334 m³/ton.

The virtual water content of rice (broken) that a consumer buys in the shop is about 3420 m³/ton. This is larger than the virtual water content of paddy rice as harvested from the field because of the weight loss if paddy rice is processed into broken rice. The virtual water content of some selected crop and livestock products for a number of selected countries are presented in Table 5.1.

Table 5.1. Average virtual water content of some selected products for a number of selected countries (m³/ton).

	USA	China	India	Russia	Indonesia	Australia	Brazil	Japan	Mexico	Italy	Nether-lands	World average*
Rice (paddy)	1275	1321	2850	2401	2150	1022	3082	1221	2182	1679		2291
Rice (husked)	1656	1716	3702	3118	2793	1327	4003	1586	2834	2180		2975
Rice (broken)	1903	1972	4254	3584	3209	1525	4600	1822	3257	2506		3419
Wheat	849	690	1654	2375		1588	1616	734	1066	2421	619	1334
Maize	489	801	1937	1397	1285	744	1180	1493	1744	530	408	909
Soybeans	1869	2617	4124	3933	2030	2106	1076	2326	3177	1506		1789
Sugar cane	103	117	159		164	141	155	120	171			175
Cotton seed	2535	1419	8264		4453	1887	2777		2127			3644
Cotton lint	5733	3210	18694		10072	4268	6281		4812			8242
Barley	702	848	1966	2359		1425	1373	697	2120	1822	718	1388
Sorghum	782	863	4053	2382		1081	1609		1212	582		2853
Coconuts		749	2255		2071		1590		1954			2545
Millet	2143	1863	3269	2892		1951		3100	4534			4596
Coffee (green)	4864	6290	12180		17665		13972		28119			17373
Coffee (roasted)	5790	7488	14500		21030		16633		33475			20682
Tea (made)		11110	7002	3002	9474		6592	4940				9205
Beef	13193	12560	16482	21028	14818	17112	16961	11019	37762	21167	11681	15497
Pork	3946	2211	4397	6947	3938	5909	4818	4962	6559	6377	3790	4856
Goat meat	3082	3994	5187	5290	4543	3839	4175	2560	10252	4180	2791	4043
Sheep meat	5977	5202	6692	7621	5956	6947	6267	3571	16878	7572	5298	6143
Chicken meat	2389	3652	7736	5763	5549	2914	3913	2977	5013	2198	2222	3918
Eggs	1510	3550	7531	4919	5400	1844	3337	1884	4277	1389	1404	3340
Milk	695	1000	1369	1345	1143	915	1001	812	2382	861	641	990
Milk powder	3234	4648	6368	6253	5317	4255	4654	3774	11077	4005	2982	4602
Cheese	3457	4963	6793	6671	5675	4544	4969	4032	11805	4278	3190	4914
Leather (bovine)	14190	13513	17710	22575	15929	18384	18222	11864	40482	22724	12572	16656

* For the primary crops, world averages have been calculated as the ratio of the global water use for the production of a crop to the global production volume. For processed products, the global averages have been calculated as the ratio of the global virtual water trade volume to the global product trade volume.

In general, livestock products have higher virtual water content than crop products. This is because a live animal consumes a lot of feed crops, drinking water and service water in its lifetime before it produces some output. We consider here an example of beef produced in an industrial farming system. It takes in average 3 years before it is slaughtered to produce about 200 kg of boneless beef. It consumes nearly 1300 kg of grains (wheat, oats, barley, corn, dry peas, soybean meal and other small grains), 7200 kg of roughages (pasture, dry hay, silage and other roughages), 24 cubic meter of water for drinking and 7 cubic meter of water for servicing. This means that to produce one kilogram of boneless beef, we use about 6.5 kg of grain, 36 kg of roughages, and 155 litres of water (only for drinking and servicing). Producing the volume of feed requires about 15340 litres of water in average.

With every step of food processing we loose part of the material as a result of selection and inefficiencies. The higher we go up in the product chain, the higher will be the virtual water content of the product. For example, the global average virtual water content of maize, wheat and rice (husked) is 900, 1300 and 3000 m^3/ton respectively, whereas the virtual water content of chicken meat, pork and beef is 3900, 4900 and 15500 m^3/ton respectively. However, the virtual water content of products strongly varies from place to place, depending upon the climate, technology adopted for farming and corresponding yields.

The units used so far to express the virtual water content of various products are in terms of cubic meters of water per ton of the product. A consumer might be more interested to know how much water it consumes per unit of consumption. One cup of coffee requires for instance 140 litres of water in average, one hamburger 2400 litres and one cotton T-shirt 2000 litres (Table 5.2).

The global average virtual water content of industrial products is 80 litres per US$. In the USA, industrial products take nearly 100 litres per US$. In Germany and the Netherlands, average virtual water content of industrial products is about 50 litres per US$. Industrial products from Japan, Australia and Canada take only 10-15 litres per US$. In world's largest developing nations, China and India, the average virtual water content of industrial products is 20-25 litres per US$.

Table 5.2. Global average virtual water content of some selected products, per unit of product.

Product	Virtual water content (litres)	Product	Virtual water content (litres)
1 glass of beer (250 ml)	75	1 glass of wine (125 ml)	120
1 glass of milk (200 ml)	200	1 glass of apple juice (200 ml)	190
1 cup of coffee (125 ml)	140	1 glass of orange juice (200 ml)	170
1 cup of tea (250 ml)	35	1 bag of potato crisps (200 g)	185
1 slice of bread (30 g)	40	1 egg (40 g)	135
1 slice of bread (30 g) with cheese (10 g)	90	1 hamburger (150 g)	2400
1 potato (100 g)	25	1 tomato (70 g)	13
1 apple (100 g)	70	1 orange (100 g)	50
1 cotton T-shirt (250 g)	2000	1 pair of shoes (bovine leather)	8000
1 sheet of A4-paper (80 g/m^2)	10	1 microchip (2 g)	32

Water footprints of nations

The global water footprint is 7450 Gm³/yr, which is 1240 m³/cap/yr in average. In absolute terms, India is the country with the largest footprint in the world, with a total footprint of 987 Gm³/yr. Though India contributes 17% to the global population, the people in India contribute only 13% to the global water footprint. On a relative basis, it is the people of the USA that have the largest water footprint, with 2480 m³/yr per capita, followed by the people in south European countries such as Greece, Italy and Spain (2300-2400 m³/yr per capita). High water footprints can also be found in Malaysia and Thailand. At the other side of the scale, the Chinese people have a relatively low water footprint with an average of 700 m³/yr per capita. The average per capita water footprints of nations are shown in Figure 5.2. The data are shown in Table 5.3 for a few selected countries.

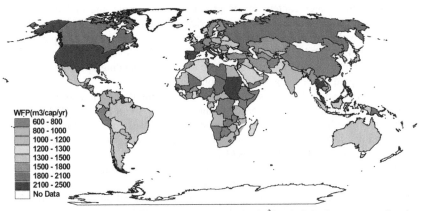

Figure 5.2. Average national water footprint per capita (m³/capita/yr). Green means that the nation's water footprint is equal to or smaller than global average. Countries with red have a water footprint beyond the global average.

The size of the global water footprint is largely determined by the consumption of food and other agricultural products (Figure 5.3). The estimated contribution of agriculture to the total water use (6390 Gm³/yr) is even bigger than suggested by earlier statistics due to the inclusion of green water use (use of soil water). If we include irrigation losses, which globally add up to about 1590 Gm³/yr (Chapagain and Hoekstra, 2004), the total volume of water used in agriculture becomes 7980 Gm³/yr. About one third of this amount is blue water withdrawn for irrigation; the remaining two thirds is green water (soil water).

Table 5.3. Composition of the water footprint for some selected countries. Period: 1997-2001.

| Country | Population | Use of domestic water resources (Gm³/yr) | | | | | Use of foreign water resources (Gm³/yr) | | | Water footprint | | Water footprint by consumption category (m³/cap/yr) | | | | |
| | | Domestic water withdrawal | Crop evaporation | | Industrial water withdrawal | | For national consumption | | For re-export of imported products | Total (Gm³/yr) | Per capita (m³/cap/yr) | Domestic water | Agricultural goods | | Industrial goods | |
			For national consumption	For export	For national consumption	For export	Agricultural goods	Industrial goods				Internal water footprint	Internal water footprint	External water footprint	Internal water footprint	External water footprint
Australia	19071705	6.51	14.03	68.67	1.229	0.12	0.78	4.02	4.21	26.56	1393	341	736	41	64	211
Bangladesh	129942975	2.12	109.98	1.38	0.344	0.08	3.71	0.34	0.13	116.49	896	16	846	29	3	3
Brazil	169109675	11.76	195.29	61.01	8.666	1.63	14.76	3.11	5.20	233.59	1381	70	1155	87	51	18
Canada	30649675	8.55	30.22	52.34	11.211	20.36	7.74	5.07	22.62	62.80	2049	279	986	252	366	166
China	1257521250	33.32	711.10	21.55	81.531	45.73	49.99	7.45	5.69	883.39	702	26	565	40	65	6
Egypt	63375735	4.16	45.78	1.55	6.423	0.66	12.49	0.64	0.49	69.50	1097	66	722	197	101	10
France	58775400	6.16	47.84	34.63	15.094	12.80	30.40	10.69	31.07	110.19	1875	105	814	517	257	182
Germany	82169250	5.45	35.64	18.84	18.771	13.15	49.59	17.50	38.48	126.95	1545	66	434	604	228	213
India	1007369125	38.62	913.70	35.29	19.065	6.04	13.75	2.24	1.24	987.38	980	38	907	14	19	2
Indonesia	204920450	5.67	236.22	22.62	0.404	0.06	26.09	1.58	2.74	269.96	1317	28	1153	127	2	8
Italy	57718000	7.97	47.82	12.35	10.133	5.60	59.97	8.69	20.29	134.59	2332	138	829	1039	176	151
Japan	126741225	17.20	20.97	0.40	13.702	2.10	77.84	16.38	4.01	146.09	1153	136	165	614	108	129
Jordan	4813708	0.21	1.45	0.07	0.035	0.00	4.37	0.21	0.22	6.27	1303	44	301	908	7	43
Mexico	97291745	13.55	81.48	12.26	2.998	1.13	35.09	7.05	7.94	140.16	1441	139	837	361	31	72
Netherlands	15865250	0.44	0.50	2.51	2.562	2.20	9.30	6.61	52.84	19.40	1223	28	31	586	161	417
Pakistan	136475525	2.88	152.75	7.57	1.706	1.28	8.55	0.33	0.67	166.22	1218	21	1119	63	12	2
Russia	145878750	14.34	201.26	8.96	13.251	34.83	41.33	0.80	3.94	270.98	1858	98	1380	283	91	5
South Africa	42387403	2.43	27.32	6.05	1.123	0.40	7.18	1.42	2.10	39.47	931	57	644	169	26	33
Thailand	60487800	1.83	120.17	38.49	1.239	0.55	8.73	2.49	3.90	134.46	2223	30	1987	144	20	41
UK	58669403	2.21	12.79	3.38	6.673	1.46	34.73	16.67	12.83	73.07	1245	38	218	592	114	284
USA	280343325	60.80	334.24	138.96	170.777	44.72	74.91	55.29	45.62	696.01	2483	217	1192	267	609	197
Global total / avg.	5994251631	344	5434	957	170.777	240	957	240	427	7452	1243	57	907	160	79	40

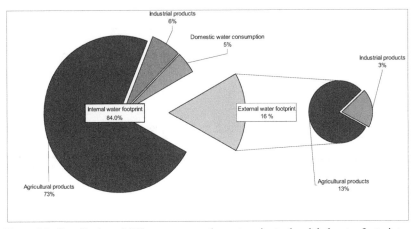

Figure 5.3. Contribution of different consumption categories to the global water footprint, with a distinction between the internal and external footprint.

The four major direct factors determining the water footprint of a country are: volume of consumption (related to the gross national income); consumption pattern (e.g. high versus low meat consumption); climate (growth conditions); and agricultural practice (water use efficiency). In rich countries, people generally consume more goods and services, meaning increased water footprints. But it is not consumption volume alone that determines the water demand of people. The composition of the consumption package is relevant too, because some goods in particular require a lot of water (bovine meat, rice). In many poor countries it is a combination of unfavourable climatic conditions (high evaporative demand) and bad agricultural practice (resulting in low water productivity) that contributes to a high water footprint. Underlying factors that contribute to bad agricultural practice and thus high water footprints are the lack of proper water pricing, the presence of subsidies, the use of water inefficient technology and lack of awareness of simple water saving measures among farmers.

The influence of the various determinants varies from country to country. The water footprint of the USA is high (2480 m^3/cap/yr) partly because of large meat consumption per capita and high consumption of industrial products. The water footprint of Iran is relatively high (1624 m^3/cap/yr) partly because of low yields in crop production and partly because of high evapotranspiration. In the USA the industrial component of the water footprint is 806 m^3/cap/yr whereas in Iran it is only 24 m^3/cap/yr.

The aggregated external water footprints of nations in the world constitute 16% of the total global water footprint (Figure 5.3). However, the share of the external water footprint strongly varies from country to country. Some African countries, such as Sudan, Mali, Nigeria, Ethiopia, Malawi and Chad have hardly any external water footprint, simply because they have little import. Some European countries on the other hand, e.g. Italy, Germany, the UK and the Netherlands have external water footprints contributing 50-80% to the total water footprint. The agricultural products that contribute most to the external water footprints of nations are: bovine meat, soybean, wheat, cocoa, rice, cotton and maize.

Eight countries – India, China, the USA, the Russian Federation, Indonesia, Nigeria, Brazil and Pakistan – together contribute fifty percent to the total global water footprint. India (13%), China (12%) and the USA (9%) are the largest consumers of the global water resources (Figure 5.4).

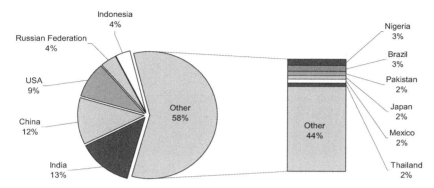

Figure 5.4. Contribution of major consumers to the global water footprint.

Both the size of the national water footprint and its composition differs between countries (Fig. 5.5). On the one end we see China with a relatively low water footprint per capita, and on the other end the USA. In the rich countries consumption of industrial goods has a relatively large contribution to the total water footprint if compared with developing countries. The water footprints of the USA, China, India and Japan are presented in more detail in Figures 5.6-5.9. The contribution of the external water footprint to the total water footprint is very large in Japan if compared to the other three countries. The consumption of industrial goods very significantly contributes to the total water footprint of the USA (32%), but not in India (2%).

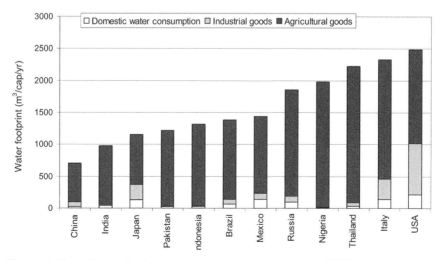

Figure 5.5. The national water footprint per capita and the contribution of different consumption categories for some selected countries.

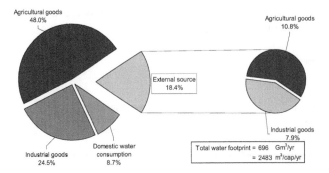

Figure 5.6. Details of the water footprints of the USA. Period: 1997-2001.

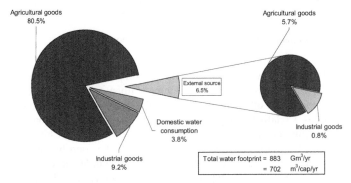

Figure 5.7. Details of the water footprints of China. Period: 1997-2001.

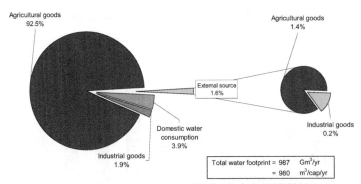

Figure 5.8. Details of the water footprints of India. Period: 1997-2001.

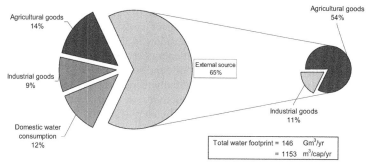

Figure 5.9. Details of the water footprints of Japan. Period: 1997-2001.

Conclusion

The global water footprint is 7450 Gm³/yr, which is in average 1240 m³/cap/yr. The differences between countries are large: the USA has an average water footprint of 2480 m³/cap/yr whereas China has an average water footprint of 700 m³/cap/yr. There are four most important direct factors explaining high water footprints. A first factor is the total volume of consumption, which is generally related to gross national income of a country. This partially explains the high water footprints of for instance the USA, Italy and Switzerland. A second factor behind a high water footprint can be that people have a water-intensive consumption pattern. Particularly high consumption of meat significantly contributes to a high water footprint. This factor partially explains the high water footprints of countries such as the USA, Canada, France, Spain, Portugal, Italy and Greece. The average meat consumption in the United States is for instance 120 kg/yr, more than three times the world-average meat consumption. Next to meat consumption, high consumption of industrial goods significantly contributes to the total water footprints of rich countries. The third factor is climate. In regions with a high evaporative demand, the water requirement per unit of crop production is relatively large. This factor partially explains the high water footprints in countries such as Senegal, Mali, Sudan, Chad, Nigeria and Syria. A fourth factor that can explain high water footprints is water-inefficient agricultural practice, which means that water productivity in terms of output per drop of water is relatively low. This factor partly explains the high water footprints of countries such as Thailand, Cambodia, Turkmenistan, Sudan, Mali and Nigeria. In Thailand for instance, rice yields averaged 2.5 ton/ha in the period 1997-2001, while the global average in the same period was 3.9 ton/ha.

Reducing water footprints can be done in various ways. A first way is to break the seemingly obvious link between economic growth and increased water use, for instance by adopting production techniques that require less water per unit of product. Water productivity in agriculture can be improved for instance by applying advanced techniques of rainwater harvesting and supplementary irrigation. A second way of reducing water footprints is to shift to consumptions patterns that require less water, for instance by reducing meat consumption. However, it has been debated whether this is a feasible road to go, since the world-wide trend has been that meat consumption increases rather than decreases. Probably a broader and subtler approach will be needed, where consumption patterns are influenced by pricing, awareness raising, labelling of products or introduction of other incentives that make people change their consumption behaviour. Water costs are generally not well reflected in the price of products due to the subsidies in the water sector. Besides, the general public is – although often aware of energy requirements – hardly aware of the water requirements in producing their goods and services.

A third method that can be used – not yet broadly recognized as such – is to shift production from areas with low water-productivity to areas with high water productivity, thus increasing global water use efficiency (Chapter 4). For instance, Jordan has successfully externalised its water footprint by importing wheat and rice products from the USA, which has higher water productivity than Jordan.

The water footprint of a nation is an indicator of water use in relation to the consumption volume and pattern of the people. As an aggregated indicator it shows the total water requirement of a nation, a rough measure of the impact of human consumption on the natural water environment. More information about the precise components and characteristics of the total water footprint will be needed, however, before one can make a more balanced assessment of the effects on the natural water systems. For instance, one has to look at what is blue versus green water use, because use of blue water often affects the environment more than green water use. Also it is relevant to consider the internal versus the external water footprint. Externalising the water footprint for instance means externalising the environmental impacts. Also one has to realise that some parts of the total water footprint concern use of water for which no alternative use is possible, while other parts relate to water that could have been used for other purposes with higher added value. There is a difference for instance between beef produced in extensively grazed grasslands of Botswana (use of green water without alternative use) and beef produced in an industrial livestock farm in the Netherlands (partially fed with imported irrigated feed crops).

The current study has focused on the quantification of consumptive water use, i.e. the volumes of water from groundwater, surface water and soil water that evaporate. The effect of water pollution was accounted for to a limited extent by including the (polluted) return flows in the domestic and industrial sector. The calculated water footprints thus consist of two components: consumptive water use and wastewater production. The effect of pollution has been underestimated however in the current calculations of the national water footprints, because one cubic metre of wastewater should not count for one, because it generally pollutes much more cubic metres of water after disposal (various authors have suggested a factor of ten to fifty). The impact of water pollution can be better assessed by quantifying the dilution water volumes required to dilute waste flows to such extent that the quality of the water remains below agreed water quality standards. We have shown this in a case study for the water footprints of nations related to cotton consumption (Chapter 8).

International water dependencies are substantial and are likely to increase with continued global trade liberalisation. Today, 16% of global water use is not for producing products for domestic consumption but for making products for export (Chapter 3). Considering this substantial percentage and the upward trend, we suggest that future national and regional water policy studies should include an analysis of international or interregional virtual water flows.

Chapter 6

Virtual versus real water transfers within China[1]

Over fifty years ago, Mao Zedong had a well-known saying 'water abundance in the South and scarcity in the North; if possible we can borrow a little bit water from South to North'. Research on the idea of transporting water from South to North China at a large scale started already fifty years ago. Today, final plans have been drafted and elements of the South-North Water Transfer Project are being implemented. Indeed, North China is suffering from water shortage and relies on water transfer from the South to release the water crisis. North China faces severe water scarcity – more than 40% of the annual renewable water resources are abstracted for human use. Nevertheless, nearly ten percent of the water used in agriculture is applied for producing food exported to South China. At the same time, North China, as China's breadbasket, annually exports substantial volumes of water-intensive products to South China. This creates a paradox in which huge volumes of water are being transferred from the water-rich South to the water-poor North while substantial volumes of food are being transferred from the food-sufficient North to the food-deficit South. This paradox, transfer of huge volumes of water from the water-rich south to the water-poor North versus transfer of substantial volumes of food from the food-sufficient North to the food-deficit South, is receiving increased attention, but the research in this field stagnates at the stage of rough estimation and qualitative description, partly due to the absence of appropriate methodologies to address the issue. The Chapter aims to review and quantify the volumes of virtual water flows between the regions in China and to put them in the context of water availability per region.

The water used in the production process of an agricultural or industrial product is called the 'virtual water' consumed in the product. In order to assess the virtual water flows between nations or regions, the basic approach has been to multiply the product trade volumes (ton/yr) by their associated virtual water contents. The virtual water content of crops has been estimated per crop per region using CROPWAT model of FAO (FAO, 2003c), climate data (FAO, 2003b) and crop data (FAO, 2003d). The virtual water content of livestock products has been calculated along the lines of 'production trees' that show different product levels (Chapagain and Hoekstra, 2003a). This study has focussed on the analysis of virtual water flows within China. Data on the virtual water flows between China and other nations have been taken from Chapagain and Hoekstra (2003a) and Hoekstra and Hung (2002).

Net import of food into a region (or net export from the region) is a function of regional production, stock changes and domestic utilisation. In this study we draw regional food balances using the same definitions as in the food balance sheets of the FAO (FAOSTAT, 2004). Net import into a region is thus equal to:

$$T_{net} = C_{dom} - Y_d - \Delta Y \tag{56}$$

---

[1] *Based on: Ma et al. (2005).*

where T_{net} denotes the net import of product p in importing region i in a particular year as a result of trade, $C_{,dom}$ the total domestic utilisation (by definition equivalent to total domestic supply), Y_d the domestic production of p, and ΔY the stock change of p. The net virtual water import related to trade in product p, A_{net} is equal to the net import volume of product p multiplied by its virtual water content V_e in the exporting region e (Equation 40).

The research area in this paper is China Mainland excluding Hong Kong, Macao and Taiwan Province. It consists of 31 provinces, municipal cities and autonomous regions. In line with the traditional regional delimitation, the country is first schematised into two regions, i.e. North and South. Both North and South are further divided into four sub-regions, each consisting of three to five provinces (Table 6.1).

Table 6.1. Regional delimitation

Region	Sub-region	Provinces
North China	North-central	Beijing, Tianjin,,Shanxi
	Northeast	Inner Mongolia, Liaoning, Jilin, Heilongjiang
	Huang-huai-hai	Hebei, Henan, Shandong, Aihui
	Northwest	Shannxi,, Gansu, Qinghai, Ningxia, Xinjiang
South China	Southeast	Shanghai, Zhejiang, Fujian
	Yangtze - area of middle and lower reaches of Yangtzi River	Jiangsu, Hubei, Hunan, Jiangxi
	South-central	Guangdong, Guanxi, Hainan
	Southwest	Chongqing, Sichuan, Guizhou, Yunnan, Tibet

As 70 percent of the total water withdrawal is used in agriculture, we focus in this study on agricultural products. We have classified these products into six categories: 'grain', 'vegetable', 'fruit', 'meat and poultry products' (further called 'meat'), 'egg and related products' (further called 'egg') and 'milk and dairy products' (further called 'dairy product'). The analysis has been carried out with data for the year 1999, when China experienced a normal hydrological year, but a good year in terms of harvest. The following assumptions have been made in this Chapter:

-    No change in product storage over the year.
-    Agricultural products imported from outside China go to the provinces with production deficits and are distributed in proportion to the deficit per province.
-    Agricultural products exported from China to other nations come from provinces with production surplus.
-    After accounting for international trade, the sub-regions with deficits import agricultural products from the closest neighbouring sub-regions with surplus.

Virtual water flows within China

Virtual water content per product category per region
The virtual water content is first calculated separately for 25 kinds of crops, 6 kinds of meats, and for eggs and milk. Calculations are done per province. Consequently, the average virtual water content for each of the six categories of agricultural products is calculated based on a production-weighted average of the virtual water content of the various products per category. The results are summarized in Table

6.2. In general the virtual water content of products in North China is higher than in South China.

North China except part of Northeast is arid or semi-arid area. The value of aridity index (the ratio of potential evaporation to precipitation) is larger than 3 or even larger than 7 in some area in Northwest. On the contrary, South China is humid or semi-humid area. Abundant sunshine and strong evaporation in North with the similar yield as South causes the virtual water content of product in North China is higher than in South China.

Table 6.2. Virtual water content of six agricultural product categories by region (m³/ton).

Region	Grain	Vegetable	Fruit	Meat, poultry and related products	Egg and related products	Milk and dairy products
North	1070	110	1060	7820	4170	1920
South	890	140	840	5670	4250	1880
National average	990	120	960	6700	4190	1910

Food trade within China
South China was the country's 'breadbasket' in the past. There was a saying "After harvest both in Hunan and Hubei, the whole country will have sufficiency". However, this situation changed since the early 1990s. In 1999 North China produced 53% of national grain, 57% vegetable, 55% fruit, 48% meat, 71% egg and 82% dairy product. On the contrary, the consumption in South China of those agricultural products all exceeded 50% of the national total.

What was it that caused the change of breadbasket from South China to North China? First of all, South China especially the South-central and Southeast where the Reform & Open Policy were initially carried out is populated densely and richest area in China. Huge investment caused prosperous of manufacturing and infrastructure construction. These occupied substantial fertile land and incentive labour shifting from agriculture to secondary and tertiary industries. Secondly, the diet change with living standard improvement needs to consume more agricultural products. Huge demand and benefit stimulate the farmer's enthusiasm in the area such as Northeast, Huang-huai-hai, where the fertile land, sunshine and heat are relatively good to extend food production. Also the national food policy-keeping food self-sufficiency at high level determines a new breadbasket should be established to substitute the old one to feed its huge population. This shifting also forms the situation that the virtual water and real water inversely flow between North and South.

At a national level China can currently realize a balance between food production and demand. However, at regional level, North China is a food surplus area and South China is a food deficit area. As to the eight sub-regions, the food surplus areas include Northeast and Huang-huai-hai in North China and Yangtze in South China. The other five regions have food deficits, of which in some densely populated and developed areas such as North-central, Southeast and South-central, the deficit accounted for more than 20% of their total demand.

Chapagain and Hoekstra (2003b) showed that in 1999 China had a net import of grain and dairy products, and a net export of vegetable, fruit and meat. Subtracting international trade South China imported 17 million tons of grain, 23 million tons of vegetable, 0.6 million tons of fruit, 1.8 million tons of meat, 2.3 million tons of egg and 2.4 million tons of dairy product from North China (Table 6.3).

Table 6.3. Food trade in China in 1999 (10^6 ton).

		Grain	Vegetable	Fruit	Meat	Egg	Dairy product
Net import from other nations[1]	North	-4.2	-2.3	-0.3	-0.2	-0.01	0.00
	South	4.5	-0.1	0.1	0.1	0.00	0.12
	National total	0.2	-2.4	-0.2	-0.1	-0.01	0.12
Net import from other region	North	-17.1	-23.2	-0.6	-1.8	-2.3	-2.4
	South	17.1	23.2	0.6	1.8	2.3	2.4

[1] Based on Chapagain and Hoekstra (2003b).

Virtual water flow between regions in China

China as a whole had a positive virtual water balance, with a net import of virtual water in 1999 of 9 billion m^3, which means around 7 m^3 per capita. The gross import was 28 billion m^3 with most virtual water going to the South. The gross export was 19 billion m^3 with the major share originating from the water-scarce North. The virtual water flow from North to South was around 52 billion m^3 (Table 6.4). The various virtual water flows between the eight sub-regions are shown in Figure 6.1.

Table 6.4. Virtual water imports and exports by region (Gm^3).

	Virtual water import from other regions within China			Virtual water import from outside China			Overall net virtual water import	
	Gross import from North	Gross import from South	Net import	Gross import	Gross export	Net import	National	Per capita
North	-	-51.6	-51.6	8.4	16.2	-7.8	-59.4	-102
South	51.6	-	51.6	19.7	2.7	17.0	68.6	104
National sum			0	28.1	18.9	9.2	9.2	7

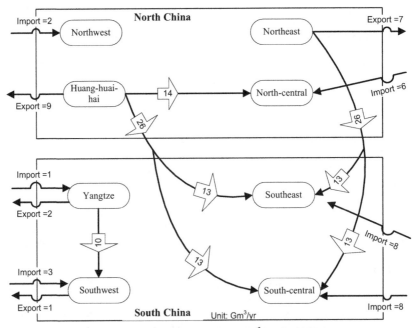

Figure 6.1. Virtual water interrelations in China (Gm^3/yr) in 1999.

Water footprint and water self-sufficiency by region

The water footprint of each regions of China has been calculated as the sum of domestic water use and the net virtual water import (Equation 44). Furthermore, the index of water self sufficiency, W_s, is also calculated as the ratio of total water use from domestic resources to the water footprint of each region. For the ease of calculation it is assumed that the imported products are entirely consumed within a region without any re-export.

In 1999 China had a water footprint of 1304 billion m^3/yr. North and South China had an equal share (Table 6.5). The average water footprint per capita in China was 1049 m^3/yr. This is lower than the global average water footprint calculated by different authors such as, 1240 m^3/cap/yr (Chapagain and Hoekstra, 2004) 900 m^3/yr per capita (according to Rockström and Gordon (2001), total global water use is 5400 Gm^3/yr, while the world population amounts to 6 billion). However in a more detailed study, by Chapagain and Hoekstra (2004) the average water footprint of an average Chinese citizen is only 702 m^3/cap/yr for the period 1997-2001. It is however much lower than the water footprint per capita in countries such as the USA, Italy and Spain, where footprints per capita are more than two to three times higher.

The people in North China transform 707 Gm^3/yr of their real water budget of 2115 Gm^3/yr into virtual water (Figure 6.2). Together with the import of virtual water from abroad, the annual virtual water budget of North China amounts to 715 Gm^3/yr. The part of this virtual water budget which is consumed within North China, 647 Gm^3/yr, constitutes North China's water footprint. The remainder, 68 Gm^3/yr is being exported to South China and abroad. South China transforms only 588 Gm^3/yr of their real water budget of 4073 Gm^3/yr into virtual water. They achieved a water footprint of 657 Gm^3/yr by having substantial virtual water imports from North China and abroad.

Table 6.5. Water footprint by region in 1999.

Region	Use of domestic water resources (Gm^3/yr)					Net virtual water import (Gm^3/yr)	Water footprint (Gm^3/yr)	Water footprint per capita (m^3/yr)	Water self-suffici-ency (%)
	Blue water			Green water	Total				
	Surface	Ground	Subtotal						
North-central	5.5	7.0	12.5	26.6	39.1	20.4	59.5	1092	66
Huang-huai-hai	56.4	33.9	90.3	255.4	345.7	-48.7	297	954	100
Northeast	49.1	23.0	72.1	122.9	195.0	-32.3	162.7	1253	100
Northwest	68.7	12.4	81.1	45.6	126.7	1.2	127.9	1425	99
North	179.7	76.3	256.0	450.5	706.5	-59.4	647.1	1106	100
Yangtze	118.1	5.9	124.0	116.6	240.6	-10.7	229.9	960	100
Southeast	48.4	1.1	49.5	25.9	75.4	33.5	108.9	1176	69
South-central	74.2	3.7	77.9	41.2	119.1	33.8	152.9	1199	78
Southwest	49.4	2.9	52.3	101.1	153.4	12.0	165.4	837	93
South	290.1	13.6	303.7	284.8	588.5	68.6	657.1	999	90
National sum	469.8	89.9	559.7	735.3	1295.0	9.2	1304.2	1049	99

At national level China had very high water self-sufficiency, with 99% in 1999. At regional level, the water-scarce North had a 100% of self-sufficiency, whereas the water-rich South relied on virtual water import, having a water self-sufficiency of only 90%. From blue water constitution we can see the ground water accounted for 30% in North China, which already approached to 95% of its ground water storage. Figure 6.2 also shows the situation that the ground water in North China was heavily exploited.

China experienced a normal year in 1999. Naturally the water balance in Figure 6.2 is based on multi-annual average value instead of the value during the dry period. Although the severe water crisis went through North China since 1990's, a lot of rivers located in remote area were still virgin territories even now, of which water accounted for huge share of annual renewable water availability. The paper only focuses on the quantitative balance of water. If the water quality is also taken into account, water that can be balanced should be deducted. Furthermore, the paper assumed the return flow reached to the river channel directly and neglected recharge to ground water. Therefore, the river discharge, 519 Gm^3/yr, looks far too optimistic that the runoff of river in North China descended approximately 20-50%, since 1990s and the drying up of river looked more and more serious in some articles such as Brown and Halweil (1998), Wang (2001), and Suo (2004) etc.

Virtual water imports and exports in relation to water availability

With further analysis of net virtual water import per sub-region, one interestingly finds that the higher per capita water availability in a sub-region, the larger the volume of virtual water import (Table 6.6). Huang-huai-hai, for instance, has a population of 310 million and a water availability of 550 m^3/person/yr, which is even less than in the Middle East and North Africa, where 300 million people occupy the limited water with a per capita share of 900 m^3/yr (Berkoff, 2003). Nevertheless, virtual water export from this region, which is regarded as one of the most water-scarce territories in the world, is quite substantial. Huang-huai-hai exported 49 Gm^3 water in virtual form in 1999, which is 157 m^3/person /yr.

What is behind the phenomena? As water is of vital importance to agriculture, logically, just from the perspective of water, virtual water export should be proportional to the water availability. In China one can find the reverse situation. Apparently other factors than water – probably availability of fertile land in particular – have been determinants in the process which has led to the current situation. Even today, the approach is mainly supply-oriented. Although concepts of demand management have long been promoted, they are in practice hardly applied.

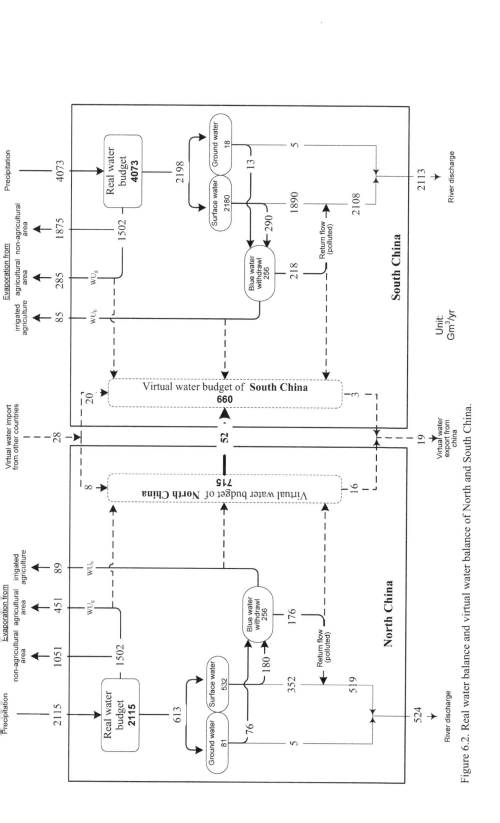

Figure 6.2. Real water balance and virtual water balance of North and South China.

Table 6.6. Virtual water import and water availability of sub- regions in China in 1999.

	Net virtual water import from within China (Gm³/yr)	Net virtual water import from outside China (Gm³/yr)	Net virtual water import (Gm³/yr)	Net virtual water import per capita (m³/yr)	Water availability per capita (m³/yr)
North-central	14.1	6.3	20.4	375	369
Huang-huai-hai	-39.9	-8.8	-48.7	-157	532
Northeast	-25.8	-6.5	-32.3	-249	1568
Northwest	0.0	1.2	1.2	14	2487
North	-51.6	7.8	-59.4	-102	1047
Yangtze	-10.2	-0.5	-10.7	-45	1821
Southeast	25.8	7.7	33.5	361	2976
South-central	25.8	8.0	33.8	265	3143
Southwest	10.2	1.8	12.0	61	5496
South	51.6	17.0	68.6	104	3347
National total	0	9.2	9.2	7	2234

The Water Transfer Project from South to North is the biggest inter-basin water transfer in the world. After a 50 year study three water diverting routes have been worked out, i.e. the west route, the middle route and the east route. Three of them will divert water from the upper, middle, and lower reaches of the Yangtze River respectively, with a maximum transfer amount of 38-43 Gm³/yr (east route 15 Gm³, middle route 13 Gm³, west route 10-15 Gm³) to meet the developing requirements of Northwest and North China (Qian et al., 2002). Until now the east and middle routes have already been carried out, covering 7 provinces and municipal cities which are Beijing, Tianjin, Hebei, Henan, Shandong, Anhui and Jiansu.

In 1999 South China imported 52 Gm³ virtual water from North China. This was more than the maximum water transfer volume by the three routes of the South-North Water Transfer Project. Furthermore, Huang-huai-hai, recipient area of the east and middle routes, had a virtual water export 26 Gm³ to South China. Although the maximum transfer amount by the two routes is 28 Gm³, it also includes the water supply to other provinces in different sub-region such as Beijing, Tian and Jiangsu. If the water going to these other provinces is subtracted, the actual water supply to Huang-huai-hai will be much less than its virtual water export.

Conclusion

Inter-basin water transfer can be realized either by real water transfers through massive engineering projects or by virtual water transfers in the form of commodities trade. In 1999, South China imported 52 Gm³ of virtual water condensed in agricultural products from North China. The year 1999 was a normal year from water resources perspective and a relatively good year from harvest point of view. The virtual water volume exceeded the planned real water transfer volume of 38-43 billion m³ per year as planning in the three South to North Water Transfer Projects.

The big question remains: is it worth the social and environmental effects bringing the water from South to North in order to bring it back in virtual form?

From a water resources point of view this doesn't make sense. There must be other decisive factors to justify the strategy. Factors that could play a role are availability of suitable cropland, possibly labour availability or national food security. A broader, integrated study would be required to give a more comprehensive assessment of the efficiency and sustainability of the South-North Water Transfer Projects.

Chapter 7

The water footprint of coffee and tea consumption[1]

Growing environmental awareness has made people more and more often ask the question: what are the hidden natural resources in a product? Which and how many natural resources were needed in order to enable us to consume a certain product? Coffee and tea consumption is possible through the use of natural and human resources in the producing countries. One of the natural resources required is water. There is a particular water need for growing the plant, but there is also a need for water to process the crop into the final product.

When there is a transfer of a product from one place to another, there is little direct physical transfer of water (apart from the water content of the product, which is quite insignificant in terms of volume). There is however a significant transfer of virtual water. In this way the coffee and tea producing countries export immense volumes of 'virtual water' to the consumer countries. Thus the importing countries are indirectly employing the water in the producing countries. As a result, consumers generally have little idea of the resources needed to enable them to consume. This Chapter aims to assess the volume of water needed to have the Dutch drink coffee and tea, in order to have concrete figures for creating awareness.

The roots of coffee consumption are probably in Ethiopia. The coffee tree is said to originate in the province of Kaffa (ICO, 2003). Coffee spread to the different parts of the world in the 17th and 18th century, the period of colonisation. Early 18th century the Dutch colonies had become the main suppliers of coffee to Europe. Today people drink coffee all over the world. The importance of coffee to people cannot easily be overestimated. Coffee is of great economic importance to the producing, mostly developing countries, and of considerable social importance to the consuming countries. Coffee is, in dollar terms, the most important agricultural product traded in the world (Dubois, 2001).

Tea is the dried leaf of the tea plant. The two main varieties of the tea plant are Camellia Sinensis and Camellia Assamica. Indigenous to both China and India, the plant is now grown in many countries around the world. Tea was first consumed as a beverage in China sometime between 2700 BC and 220 AD (L'Amyx, 2003). The now traditional styles of green, black and oolong teas first made an appearance in the Ming Dynasty in China (1368-1644 AD). Tea began to travel as a trade item as early as the fifth century with some sources indicating Turkish traders bartering for tea on the Mongolian and Tibetan borders. Tea made its way to Japan late in the sixth century, along with another famous Chinese export product - Buddhism. By the end of the seventh century, Buddhist monks were planting tea in Japan. Tea first

---

[1] Based on: Chapagain and Hoekstra (2003c; 2003d); Hoekstra and Chapagain (2004a); Chapagain and Hoekstra (submitted-b) as manuscript for publication in Ecologial Economics.

arrived in the west via overland trade into Russia. Certainly Arab traders had dealt in tea prior to this time, but no Europeans had a hand in tea as a trade item until the Dutch began an active and lucrative trade early in the seventeenth century. Dutch and Portuguese traders were the first to introduce Chinese tea to Europe. The Portuguese shipped it from the Chinese coastal port of Macao; the Dutch brought it to Europe via Indonesia (Twinings, 2003b). From Holland, tea spread relatively quickly throughout Europe (L'Amyx, 2003). Although drunk in varying amounts and different forms, tea is the most consumed beverage in the world next to water (Sciona, 2003). Tea is grown in over 45 countries around the world, typically between the Tropics of Cancer and Capricorn (FAO, 2003d). The study is limited to tea made from the real tea plant, of which the two main varieties are Camellia Sinensis and Camellia Assamica. This excludes other sorts of 'tea', made from other plants, such as 'rooibos tea' (from a reddish plant grown in South Africa), 'honeybush tea' (related to rooibos tea and also grown in South Africa), 'yerba mate' (from a shrub grown in some Latin American countries), and 'herbal tea' (a catch-all term for drinks made from leaves or flowers from various plants infused in hot water).

As a first step, we estimate the virtual water content of coffee and tea in each of the countries that export coffee or tea to the Netherlands. Next, we calculate the volumes of virtual water flows entering and departing the Netherlands in the period 1995-99 insofar as they are related to coffee and tea trade. Finally, we assess the volume of water needed to drink one cup of coffee or tea in the Netherlands. The water volume per cup multiplied with the number of cups consumed per year provides an estimate of the total Dutch annual water footprint related to coffee or tea consumption.

Virtual water content of coffee and tea

The virtual water content of coffee or tea is the volume of water required to produce one unit of coffee or tea, generally expressed in terms of cubic metres of water per ton of coffee or tea. This is different in the different stages of processing. In the case of coffee, the virtual water content of fresh cherries is calculated based on the crop water requirement of the coffee plant (in m^3/ha) and the yield of fresh cherries (in ton/ha). After each processing step, the weight of the remaining product is smaller than the original weight. The virtual water content of coffee and tea products are estimated following the methodology proposed by Chapagain and Hoekstra (2003a). The virtual water content of the resulting product (expressed in m^3/ton) is larger than the virtual water content of the original product. It can be found by dividing the virtual water content of the original product by the product fraction. If a particular processing step requires water (viz. the processes of pulping, fermentation and washing in the wet production method), the water needed (in m^3 per ton of original product) is added to the initial virtual water content of the original product before translating it into a value for the virtual water content of the resulting product.

Figure 7.1 shows how the virtual water content of coffee is calculated in its subsequent production stages in both the case of the wet production method and the case of the dry production method. The numbers are based on the example of Brazil. This scheme of calculation is adopted for all other countries with their respective crop water requirement and the yield of fresh cherry. The product fractions and process water requirements are assumed to be constant across different coffee producing countries.

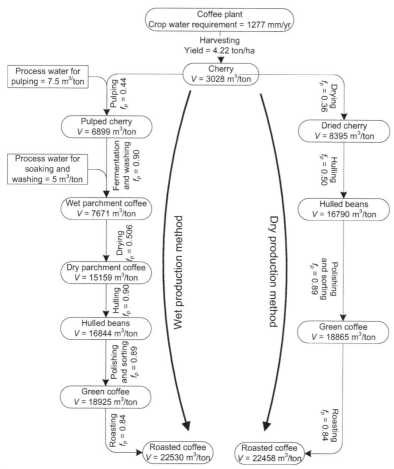

Figure 7.1. Steps in the calculation of the virtual water content of coffee under the different production methods. The numbers are for Brazil. f_p is the product fraction (ton/ton of primary product and V is the virtual water content (m³/ton).

In the case of tea, the virtual water content of fresh leaves is calculated based on the crop water requirement of the tea plant (in m³/ha) and the yield of fresh leaves (in ton/ha). The virtual water content at different stages of production is calculated using the same approach followed for coffee. Figure 7.2 shows how the virtual water content of tea is calculated in its subsequent production stages in the case of the orthodox production of black tea in India.

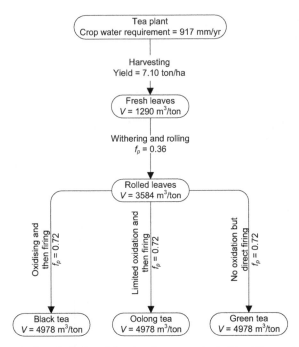

Figure 7.2. Steps in the calculation of the virtual water content of tea. The numbers are for
India. f_p is the product fraction (ton/ton of primary product and V is the virtual
water content (m^3/ton).

The annual crop water requirements of a coffee or tea plant are calculated per
country using the CROPWAT model developed by the Food and Agriculture
Organization (FAO, 2003c). The crop coefficients for the crops have been taken
from Allen *et al.* (1998). The climate data required as input into the CROPWAT
model have been taken from the CLIMWAT database (FAO, 2003b). In the cases
where this database contains data for a number of climate stations within a country,
the data from the station in the capital has been adopted. It is admitted that this is a
crude assumption, because the climate near the capital is not necessarily
representative for the climate in the areas in the country where the specific crops are
grown, but global data on exact locations of the plantations area are not easily
obtainable. Country-specific data on crop production per unit of land (ton/ha) have
been obtained from the FAOSTAT database (FAO, 2003d). The figures provided in
the database refer to yields in terms of green coffee and 'made tea'. Yields in terms
of fresh cherries have been calculated based on the ratio of green coffee weight to
fresh cherry weight. Yields in terms of fresh tea-leaves have been calculated based
on the ratio of made tea weight to fresh leaves weight.

From fresh cherries to green coffee the weight is reduced to about 16 per cent of
the original weight, due to removing pulp and parchment, reduction in moisture
content and sorting out of low-quality beans (GTZ, 2002c). The weight reduction
occurs in steps. In the wet production method, only 44% of the fresh cherry remains
after pulping (Bressani, 2003), 90% of the pulped cherry remains after fermentation
and washing (Bressani, 2003), 51% of the wet parchment coffee remains after
drying (GTZ, 2002a) and 80% of the dry parchment coffee remains after hulling,

polishing and sorting (GTZ, 2002c). In the dry production method, about 36% of the fresh cherry remains after drying (Hicks, 2001), 50% of the dried cherry remains after hulling (Hicks, 2001) and 89% of the hulled beans remains after polishing and sorting. From green coffee to roasted coffee there is another weight reduction, due to reduction in moisture content. The remaining fraction after roasting is generally reported to be 84% of the original green coffee (GTZ, 2002c; Hicks, 2001; ICO, 2003; Sovrana, 2003).

The wet production method requires water for pulping, fermentation and washing process. The total amount of water needed ranges between 1 and 15 m^3 per ton of cherry (GTZ, 2002b). In this study we crudely assume that 7.5 m^3 of water per ton of fresh cherry is needed in the pulping process and that 5 of water per ton of pulped cherry is needed in the fermentation and washing process (Roast and Post, 2003). If we bring these two numbers into one denominator, this is equivalent to about 10 m^3 of water per ton of cherry. We will see later that the overall result of the study, the estimated total water needs for making coffee, are not sensitive to the assumptions made here.

Table 7.1 shows the calculations for all coffee producing countries from which the Netherlands imports coffee for the wet production method. These countries together are responsible for 84 per cent of the global coffee production. The differences between the two production methods in terms of total water needs are very small. The virtual water content of green coffee is 17.63 m^3/kg for the wet production method, whereas it is 17.57 m^3/kg for the dry production method (global averages). The water needs for roasted coffee are 20.98 and 20.92 m^3/kg respectively. Most water is needed for growing the coffee plant. In the wet production method, only 0.34% of the total water need refers to process water.

According to Duke (1983), 10 kg of green shoots (containing 75-80% water) produce about 2.5 kg of dried tea. The overall remaining fraction after processing fresh tealeaves into made tea is thus 0.25. The weight reduction occurs in two steps. Withering reduces moisture content up to 70% and drying further reduces it down to about 3% (Twinings, 2003a). There is no reduction of weight in the rolling and oxidation processes. Due to the higher firing temperatures, oolong teas contain less moisture and have a longer shelf life than green teas. In this study, the remaining fraction after withering is taken as 0.72 (ton of withered tea per ton of fresh leaves) and a remaining fraction after firing is taken as 0.36 (ton of black tea per ton of rolled leaves). The different methods of processing fresh tea-leaves into black, green or oolong tea are more or less equal if it comes to the remaining fraction after all (ton of made tea per ton of fresh tealeaves). When calculating the virtual water content of tea in the different tea-producing countries no distinction is made between different production methods. The calculation is made for black tea and assumed to represent for green tea and oolong tea as well.

Table 7.1. Virtual water content of coffee produced with the wet production method by country. Period 1995-99.

	Production of green coffee (ton/yr)	Yield of green coffee (ton/ha)	Yield of fresh cherry (ton/ha)	Crop water requirement (mm/yr)	Virtual water content (m³/ton)						
					Fresh cherry	Pulped cherry	Wet parchment coffee	Dry parchment coffee	Hulled beans	Green coffee	Roasted coffee
Brazil	1370232	0.68	4.22	1277	3028	6882	7671	15159	16844	18925	22530
Colombia	689688	0.74	4.61	893	1939	4406	4920	9723	10803	12139	14451
Indonesia	466214	0.55	3.41	1455	4268	9699	10802	21347	23719	26650	31727
Vietnam	384220	1.87	11.63	938	807	1833	2061	4074	4526	5086	6054
Mexico	329297	0.46	2.88	1122	3898	8859	9868	19502	21669	24347	28985
Guatemala	240222	0.90	5.60	1338	2388	5428	6055	11967	13296	14940	17786
Uganda	229190	0.84	5.25	1440	2741	6230	6947	13729	15254	17139	20404
Ethiopia	227078	0.91	5.65	1151	2036	4628	5167	10212	11346	12749	15177
India	220200	0.81	5.08	754	1485	3375	3774	7459	8288	9312	11086
Costa Rica	157188	1.47	9.14	1227	1342	3051	3414	6748	7497	8424	10028
Honduras	154814	0.78	4.87	1483	3044	6919	7712	15241	16935	19028	22652
El Salvador	138121	0.85	5.28	1417	2685	6102	6805	13448	14942	16789	19987
Ecuador	121476	0.32	1.98	1033	5225	11875	13219	26125	29028	32616	38828
Peru	116177	0.61	3.80	994	2612	5937	6621	13084	14538	16335	19446
Thailand	75814	1.12	6.96	1556	2236	5082	5671	11208	12453	13993	16658
Venezuela	67802	0.35	2.19	1261	5756	13082	14560	28775	31972	35923	42766
Nicaragua	65373	0.73	4.55	1661	3649	8294	9240	18260	20289	22797	27139
Madagascar	63200	0.33	2.04	1164	5692	12935	14397	28453	31614	35521	42287
Tanzania	44540	0.38	2.38	1422	5964	13555	15085	29812	33125	37219	44308
Bolivia	22613	0.94	5.84	1093	1874	4258	4756	9398	10443	11733	13968
Togo	14416	0.34	2.12	1409	6643	15097	16799	33199	36887	41447	49341
Sri Lanka	11133	0.68	4.22	1426	3379	7680	8558	16913	18793	21115	25137
Panama	10726	0.41	2.55	1294	5068	11517	12822	25339	28155	31634	37660
Ghana	4909	0.35	2.16	1381	6402	14549	16190	31996	35552	39946	47554
USA	2924	1.24	7.74	938	1212	2754	3085	6097	6774	7611	9061
Average*		0.80	4.53	1195	2820	6409	7145	14121	15690	17629	20987

* Country data have been weighted on the basis of their share of green coffee to the global production, which is 6,201,976 ton/yr.

Table 7.2 presents the virtual water content of tea in different production steps for all tea producing countries that export tea to the Netherlands. These countries together are responsible for 81 per cent of the global tea production. The global average virtual water content of fresh tealeaves is 2.7 m^3/kg. The average virtual water content of made tea is 10.4 m^3/kg. The latter figure has been based on a calculation for black tea, but there would be hardly any difference for green tea or oolong tea, because the overall weight reduction in the case of green tea or oolong tea is similar to the weight reduction when producing black tea. Besides, it is good to note here that black tea takes the largest share in the global production of tea (78%).

Table 7.2. Virtual water content of tea by country. Period: 1995-99.

Countries	Production (ton/yr)	Yield (ton/ha) made tea[1]	fresh tea-leaves	Crop water requirement (mm/yr)	Virtual water content (m^3/ton) Fresh tea-leaves	Withered and rolled leaves	Made tea
Argentina	53124	1.40	5.39	1286	2387	6630	9208
Bangladesh	51912	1.08	4.15	1404	3383	9397	13052
Brazil	6753	1.84	7.11	1550	2180	6055	8410
China	649489	0.73	2.80	1205	4304	11955	16604
India	794180	1.84	7.10	917	1290	3584	4978
Indonesia	160334	1.43	5.51	1769	3213	8924	12395
Japan	87140	1.68	6.47	1165	1802	5004	6950
Mauritius	2206	2.15	8.31	1548	1864	5178	7191
South Africa	10866	1.66	6.41	1822	2842	7894	10965
Sri Lanka	269013	1.41	5.45	1731	3174	8817	12247
Tanzania	24140	1.29	4.98	1726	3467	9632	13377
Turkey	146756	1.91	7.38	1349	1828	5078	7053
Uganda	20365	1.12	4.32	1746	4046	11239	15610
Weighted mean					2694	7483	10394

[1]Source: FAO (2003d).

Virtual water flows related to the trade of coffee and tea

The volume of virtual water imported into the Netherlands (in m^3/yr) as a result of coffee or tea import can be found by multiplying the amount of product imported (in ton/yr) by the virtual water content of the product (in m^3/ton). The virtual water content of tea and coffee is taken from the exporting countries. The volume of virtual water exported from the Netherlands is calculated by multiplying the export quantity by the respective average virtual water content of coffee and tea in the Netherlands. The latter is taken as the average virtual water content of the coffee and tea imported into the Netherlands. The difference between the total virtual water import and the total virtual water export is the net virtual water import to the Netherlands, an indicator for the total amount of water needed to have the Dutch drink coffee and tea.

Data on coffee and tea trade have been taken from the database, PC-TAS, produced by the United Nations Statistics Division (UNSD) in New York in collaboration with the International Trade Centre (ITC) in Geneva for the period 1995-99 (ITC, 1999). The total volumes of coffee and tea imported into the Netherlands and the total volumes exported are presented in Table 7.3. The data are given for four different coffee products: 'non-decaffeinated non-roasted coffee', 'decaffeinated non-roasted coffee', 'non-decaffeinated roasted coffee' and 'decaffeinated roasted coffee'. The term 'non-roasted coffee' in PC-TAS refers to what is generally called 'green coffee'. Some of the countries exporting coffee to the Netherlands do not grow coffee themselves. These countries import the coffee from elsewhere in order to further trade it. The UNSD uses the different terminology as 'fermented' and 'not fermented' tea for oxidised (black) and non-oxidised (green) tea respectively.

Table 7.3. Coffee and tea import into and export from the Netherlands by product type during the period 1995-99.

Product code in PC-TAS	Product	Import (ton/yr)	Export (ton/yr)
090111	Coffee, not roasted, not decaffeinated	135381	7252
090112	Coffee, not roasted, decaffeinated	5331	731
090121	Coffee, roasted, not decaffeinated	22020	7229
090122	Coffee, roasted, decaffeinated	3887	1444
090210	Green tea (not fermented) in packages < 3 kg	225	51
090220	Green tea (not fermented) in packages > 3 kg	936	17
090230	Black tea (fermented) & partly fermented tea in packages < 3 kg	2580	1346
090240	Black tea (fermented) & partly fermented tea in packages > 3 kg	13485	7977

The virtual water import to the Netherlands as a result of coffee import is 2953 Mm³/yr (Table 7.4). Brazil and Colombia together are responsible for 25 percent of this import. Other important sources are Guatemala (5%), El Salvador (5%) and Indonesia (4%) as shown in Figure 7.3. A large part of the coffee import comes from the non-coffee-producing countries Belgium and Germany (34% in total). The virtual water import to the Netherlands as a result of tea import in the period 1995-99 has been 197 Mm³/yr in average (Table 7.5). Indonesia is the largest source (contributing 35% of the total import into the Netherlands). Other sources are China (21%), Sri Lanka (14%), Argentina (6%), India (5%), Turkey (3%) and Bangladesh (1%). There is also some import from within Europe: Germany (6%), Switzerland (4%), United Kingdom (2%) and Belgium-Luxemburg (2%). It is difficult to trace back the original source of the coffee and tea imported from the countries that do not produce the crop themselves. For the coffee and tea imported from countries that do not produce the crop themselves, the global average virtual water content of the product has been taken from Table 7.1 and Table 7.2.

Table 7.4. Average annual virtual water import to the Netherlands related to coffee import in the period 1995-99.

	Virtual water import (10^6 m³/yr)	Share of total import volume (%)
Belgium-Luxemburg	612	20.7
Brazil	426	14.4
Germany	380	12.9
Colombia	324	11.0
Guatemala	159	5.4
El Salvador	154	5.2
Indonesia	127	4.3
Togo	99	3.3
Tanzania	92	3.1
Mexico	85	2.9
Costa Rica	75	2.6
Nicaragua	73	2.5
Peru	72	2.4
Honduras	48	1.6
India	36	1.2
France	34	1.2
Uganda	32	1.1
Ecuador	19	0.6
Italy	19	0.6
Others	89	2.0
Total	2953	100

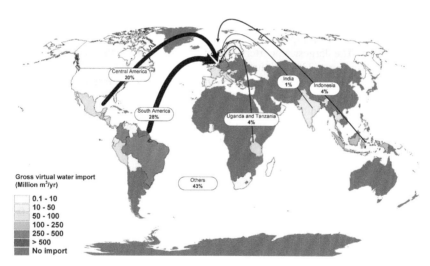

Figure 7.3. Virtual water import to the Netherlands related to coffee imports. The greener the area the more the import to the Netherlands.

The total import of green coffee over the period 1995-99 amounts to 141×10^3 ton/yr. The import of roasted coffee is 26×10^3 ton/yr. The average virtual water content of coffee imported into the Netherlands is 17.1 m³ per kg of green coffee and 20.4 m³ per kg of roasted coffee. These figures are very close to the average global virtual water content of green and roasted coffee respectively. The total import of tea over the period 1995-99 amounts to 17×10^3 ton/yr. The average virtual water content of tea imported into the Netherlands is 11.4 m³ per kg of made tea. This figure is very close to the global average virtual water content of made tea, which is 10.4 m³/ kg.

Table 7.5. Average annual virtual water import to the Netherlands related to tea import in the period 1995-99.

	Virtual water import (10^6 m³/yr)	Share of total import volume (%)
Indonesia	69.2	35.2
China	41.2	21.0
Sri Lanka	28.2	14.4
Argentina	12.0	6.1
Germany	11.5	5.8
India	8.6	4.4
Switzerland-Liecht.	7.2	3.7
Turkey	6.1	3.1
UK	3.5	1.8
Belgium-Luxembourg	3.2	1.6
Bangladesh	1.9	1.0
Tanzania	1.0	0.5
Brazil	0.9	0.5
Others	2.1	1.1
Total	197	100

The total virtual water export from the Netherlands as a result of coffee export is 314 Mm³/yr. The largest importers of virtual water from the Netherlands are: Belgium-Luxemburg (23%), United Kingdom (20%), Germany (18%), and France (12%). The total virtual water export from the Netherlands as a result of tea export is 107 Mm³/yr. The largest importers of virtual water from the Netherlands are: Germany (21%), United Kingdom (17%), the Russian Federation (15%), Switzerland (8%), USA (6%), Italy (4%), France (4%), and Belgium-Luxemburg (3%). Virtual water export from the Netherlands as a result of coffee and tea export are presented in Table 7.6 and Table 7.7 respectively.

Coffee and tea import are responsible for about 5 per cent of the total gross virtual water import into the Netherlands related to the import of agricultural products (Table 7.8). The Dutch water footprint related to tea consumption is much smaller than the Dutch water footprint related to coffee consumption, due to the fact that the Dutch consume relatively large amounts of coffee and that tea has much lower virtual water content than coffee.

Table 7.6. Average annual virtual water export from the Netherlands related to coffee export in the period 1995-99.

	Virtual water import (10^6 m^3/yr)	Share of total import volume (%)
Belgium-Luxemburg	73.5	23.4
UK	61.1	19.5
Germany	55.5	17.7
France	38.7	12.3
Sweden	16.8	5.3
Spain	10.9	3.5
Denmark	9.9	3.1
USA	7.6	2.4
Russian Federation	6.9	2.2
Italy	4.3	1.4
Norway	4.1	1.3
Finland	2.9	0.9
Netherlands-Antil.	2.0	0.6
Austria	1.6	0.5
Lithuania	1.5	0.5
Greece	1.4	0.5
Czech Republic	1.4	0.4
Aruba	1.3	0.4
Portugal	1.2	0.4
Turkey	1.0	0.3
Others	10.3	3.3
Total	314.0	100

Table 7.7. Average annual virtual water export from the Netherlands related to tea export in the period 1995-99.

Origin	Virtual water export (10^6 m^3/yr)	Share of total export volume (%)
Germany	22.2	20.7
UK	18.6	17.3
Russian Federation	16.5	15.4
Switzerland-Liecht	8.1	7.6
USA	6.6	6.2
Italy	4.5	4.2
France	4.3	4.0
Belgium-Luxembourg	2.7	2.5
Saudi Arabia	2.4	2.3
Denmark	2.2	2.1
Canada	1.6	1.5
Austria	1.5	1.4
Finland	1.4	1.3
Others	14.6	13.6
Total	107	100

Table 7.8. Virtual water imports into and exports from the Netherlands related to trade in tea, coffee, crops and livestock products.

	Gross import of virtual water (Mm³/yr)	Gross export of virtual water (Mm³/yr)	Net import of virtual water (Mm³/yr)
Related to tea trade (1995-99)	197	107	90
Related to coffee trade (1995-99)	2953	314	2639
Total (coffee and tea)	3150	421	2729
Related to trade in crops and crop products (1997-2001)*	48607	34529	14078
Related to trade in livestock and livestock products (1997-2001)*	7852	15146	-7294
Related to total trade in agricultural products	56459	49675	6784

* Source: Chapagain and Hoekstra (2004).

The water needed to drink a cup of coffee or tea

The quantity of roasted coffee per cup of coffee is not a fixed figure. It is differs among people. The Speciality Coffee Association of America (SCAA) suggests 10 gram per cup as the proper measure for brewed coffee if using the American standards. As per López-Ortiz and Owen (2003), one standard European cup of coffee contains 7 gram per cup. A figure of 5 gram per cup is recommended by Van Wieringen (2001). For the calculation of the virtual water content of a standard cup of coffee it is assumed to contain 7 gram of roasted coffee in this study. Based on the average virtual water content of roasted coffee (20.4 m³/kg), one cup of coffee requires about 140 litres of water in total. A standard cup of coffee is 125 ml, which means that we need more than 1100 drops of water for producing one drop of coffee. For making one kilogram of soluble coffee powder, one needs 2.3 kg of green coffee (Rosenblatt et al., 2003). That means that the virtual water content of instant coffee is about 39400 m³/ton. This is much higher than in the case of roasted coffee, but for making one cup of instant coffee one needs a relatively small amount of coffee powder (about 2 gram). Surprisingly, the virtual water content of a cup of instant coffee is thus lower than the virtual water content of a cup of normal coffee.

A cup of tea typically requires 1.5 to 3 grams of processed tea (The Fragrant Leaf, 2003). Here, it is assumed that 3 grams of processed tea (either black, green or oolong tea) for a cup of normal tea and 1.5 gram for 'weak' tea. This is equivalent to 34 litres of water per standard cup of tea and 17 litres per cup for weak tea. One consumes about 4 times more water if the choice is made for a cup of coffee instead of a cup of tea. With a standard cup of tea of 250 ml, about 136 drops of water for producing one drop of tea is needed. The results as presented in Table 7.9 for the Netherlands are quite representative for the global average, so the figures can be cited in more general terms as well.

Table 7.9. Virtual water content of a cup of tea or coffee.

		Virtual water content of the dry ingredient (m^3/kg)	One cup of tea or coffee		
			Dry product content (gram/cup)	Real water content (litre/cup)	Virtual water content (litre/cup)
Coffee	- Standard cup of coffee	20.4	7	0.125	140
	- Weak coffee	20.4	5	0.125	100
	- Strong coffee	20.4	10	0.125	200
	- Instant coffee	39.4	2	0.125	80
Tea	- Standard cup of tea	11.4	3	0.250	34
	- Weak tea	11.4	1.5	0.250	17

The water footprint of coffee and tea consumption

With an average of 3 cups of coffee a day per person, the Dutch community has a coffee-related water footprint of 2.6 billion cubic metres of water per year. This is the volume of water appropriated in the coffee-producing countries. The total volume is equivalent to 36% of the annual flow of the Meuse, one of the Dutch rivers. The Dutch people account for 2.4% of the world coffee consumption. All together, the world population requires about 110 billion cubic metres of water per year in order to be able to drink coffee. This is equivalent to 15 times the annual Meuse runoff, or 1.5 times the annual Rhine runoff.

The water footprint of the Dutch tea consumption is 90 million cubic metres per year. Dutch people contribute only 0.28% (7.8×10^3 ton/yr) to the world tea consumption (2.82 million ton/yr). The world population requires about 30 billion cubic metres of water per year in order to be able to drink tea.

The water needed to drink coffee or tea in the Netherlands is actually not Dutch water, because the crops are not produced in the Netherlands. Coffee is produced in Latin America (Brazil, Colombia, Guatemala, El Salvador, Mexico, Costa Rica, Nicaragua, Peru, Honduras), Africa (Togo, Tanzania, Uganda) and Asia (Indonesia, India). The most important sources are Brazil and Colombia. Tea is produced in South East Asia (Indonesia, China, Sri Lanka, India, and Bangladesh) and some other countries in different parts of the world (Argentina, Turkey, Brazil, Tanzania, and South Africa). There are also coffee and tea imports from countries that do not produce coffee or tea themselves, such as Germany and Belgium. They are merely intermediate countries, where tea is just transited or upgraded (e.g. through blending or making brand names to gain higher economic returns).

Conclusion

The consumption of coffee and tea in the Netherlands has positive impacts on the economies of the producing countries. It generates economic benefits to the producing countries (which are mostly developing countries) with the use of a resource (rainwater) that has relatively low opportunity cost. Although rainwater appropriated for coffee or tea production will often have no alternative use (e.g. production of another crop or natural forest) that will provide higher economic or social return, it is bad economic policy to completely leave out the opportunity cost or scarcity value of rainwater from the price of the product (Albersen et al., 2003;

Hoekstra *et al.*, 2001; Hoekstra *et al.*, 2003). In the exceptional cases where irrigation is applied, the opportunity costs of the water inputs are considerably higher than in the case of rainwater use. In those circumstances, it is even more important to pass on the costs of the water to the consumers of the coffee or tea, something which is thus far unusual.

The volume of water needed to make coffee and tea depends particularly on the climate at the place of production and the yields per hectare that are obtained. The latter partly depends on the climatic conditions, but also on soil conditions and management practice. For the overall water needs, it makes hardly any difference whether coffee is produced with the dry or the wet production process, because the water used in the wet production process is only a very small fraction (0.34%) of the water used to grow the coffee plant. However, this relatively small amount of water can be and actually often is a problem, because this is water to be obtained from surface or groundwater, which is generally scarcer than rainwater (i.e. competition is larger). Besides, the wastewater from the coffee factories is often heavily polluted (GTZ, 2002b). In current practice, coffee as bought by final consumers does neither include in its price the economic costs of water inputs and water impacts, nor reveal qualitative information about those costs on its label. It is necessary to further explore the economic, social and environmental relevance and practical modes of passing on the costs of water inputs and impacts in the price of coffee in the shop.

Chapter 8

The water footprint of cotton consumption[1]

Globally, freshwater resources are becoming scarcer due to an increase in population and subsequent increase in water appropriation and deterioration of water quality. The impact of consumption of people on the global water resources can be mapped with the concept of the 'water footprint', a concept introduced by Hoekstra and Hung (2002) and subsequently elaborated by Chapagain and Hoekstra (2004). The water footprint of a nation has been defined as the total volume of freshwater that is used to produce the goods and services consumed by the inhabitants of the nation. It deviates from earlier indicators of water use in the fact that the water footprint shows water demand related to *consumption* within a nation, while the earlier indicators (e.g. total water withdrawal for the various sectors of economy) show water demand in relation to *production* within a nation. The current report focuses on the assessment and analysis of the water footprints of nations insofar related to the consumption of cotton products. The period 1997-2001 has been taken as the period of analysis.

The water footprint concept is an analogue of the ecological footprint concept which was introduced in the 1990s (Rees, 1992; Wackernagel *et al.*, 1999; Wackernagel *et al.*, 1997; Wackernagel and Rees, 1996). Whereas the ecological footprint denotes the *area* (hectares) needed to sustain a population, the water footprint represents the *water volume* (cubic metres per year) required.

Earlier water-footprint studies were limited to the quantification of resource use, i.e. the use of groundwater, surface water and soil water (Chapagain and Hoekstra, 2003a; 2003c; 2003d; 2004; Hoekstra and Hung, 2002). This Chapter extends the water footprint concept through quantifying the impacts of pollution as well. This has been done by quantifying the dilution water volumes required to dilute waste flows to such extent that the quality of the water remains below agreed water quality standards. The rationale for including this water component in the definition of the water footprint is similar to the rationale for including the land area needed for uptake of anthropogenic carbon dioxide emissions in the definition of the ecological footprint. Land and water do not function as resource bases only, but as systems for waste assimilation as well. The method to translate the impacts of pollution into water requirements as applied in this study can potentially invoke a similar debate as is being held about the methods applied to translate the impacts of carbon dioxide emissions into land requirements (see e.g. Van den Bergh and Verbruggen (1999), Van Kooten and Bulte (2000)). Such a debate is always good, because of the societal need for proper natural resources accounting systems on the one hand and the difficulties in achieving the required scientific rigour in the accounting procedures on the other hand. The approach introduced in the current study should be seen as a

---

[1] *Based on: Chapagain et al. (2005c); Chapagain et al. (2005d) accepted for publication in Ecological Economics.*

first step; reflection on the possible improvements is made in the concluding Section of this Chapter.

Some of the earlier studies on the impacts of cotton production were limited to the impacts in the industrial stage only (e.g. Ren (2000)), leaving out the impacts in the agricultural stage. Other cotton impact studies use the method of life cycle analysis and thus include all stages of production, but these studies are focussed on methodology rather than the quantification of the impacts (e.g. Proto et al. (2000), Seuring, (2004)). Earlier studies that go in the direction of what we aim at in this report are the background studies for the cotton initiative of the World Wide Fund for Nature (Soth et al., 1999; De Man, 2001). In this study, however, it is aimed to synthesize the various impacts of cotton on water in one comprehensive indicator, the water footprint, and we introduce the spatial dimension by showing how water footprints of some nations particularly press in other parts of the world.

Cotton is the most important natural fibre used in the textile industries worldwide. Today, cotton takes up about 40 percent of textile production, while synthetic fibres take up about 55% (Proto et al., 2000; Soth et al., 1999). During the period 1997-2001, international trade in cotton products constitutes 2 percent of the global merchandise trade value.

The impacts of cotton production on the environment are easily visible and have different faces. On the one hand there are the effects of water depletion, on the other hand the effects on water quality. In many of the major textile processing areas, downstream riparians can see from the river what was the latest colour applied in the upstream textile industry. The Aral Sea is the most famous example of the effects of water abstractions for irrigation. In the period 1960-2000, the Aral Sea in Central Asia lost approximately 60% of its area and 80% of its volume (1998; Hall et al., 2001; Pereira et al., 2002; UNEP, 2002; WWF, 2004) as a result of the annual abstractions of water from the Amu Darya and the Syr Darya – the rivers which feed the Aral Sea – to grow cotton in the desert.

About 53 percent of the global cotton field is irrigated, producing 73 percent of the global cotton production (Soth et al., 1999). Irrigated cotton is mainly grown in the Mediterranean and other warm climatic regions, where freshwater is already in short supply. Irrigated cotton is mainly located in dry regions: Egypt, Uzbekistan, and Pakistan. The province Xinjiang of China is entirely irrigated whereas in Pakistan and the North of India a major portion of the crop water requirements of cotton are met by supplementary irrigation. As a result, in Pakistan already 31 percent of all irrigation water is drawn from ground water and in China the extensive freshwater use has caused falling water tables (Soth et al., 1999). Nearly 70 percent of the world's cotton crop production is from China, USA, India, Pakistan and Uzbekistan (USDA, 2004). Most of the cotton productions rely on a furrow irrigation system. Sprinkler and drip systems are also adopted as an irrigated method in water scarce regions. However, hardly about 0.7 percent of land in the world is irrigated by this method (Postel, 1992).

Role of green, blue and dilution water in cotton production

From field to end product, cotton passes through a number of distinct production stages with different impacts on water resources. These stages of production are often carried out at different locations and consumption can take place at yet another place. For instance, Malaysia does not grow cotton, but imports raw cotton from

China, India and Pakistan for processing in the textile industry and exports cotton clothes to the European market. For that reason the impacts of consumption of a final cotton product can only be found by tracing the origins of the product. The relation between the production stages and their impacts on the environment is shown in Figure 8.1.

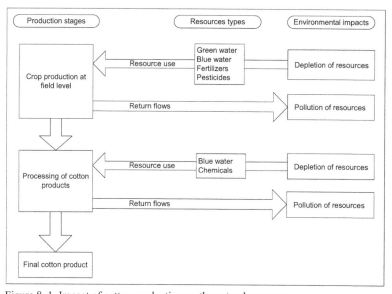

Figure 8. 1. Impact of cotton production on the natural resources.

Although the chain from cotton growth to final product can take several distinct steps, there are two major stages: the agricultural stage (cotton production at field level) and the industrial stage (processing of seed cotton into final cotton products). In the first stage, there are three types of impact: evaporation of infiltrated rainwater for cotton growth, withdrawal of ground- or surface water for irrigation, and water pollution due to the leaching of fertilisers and pesticides. Following Falkenmark (1995), the term 'green water use' is used for the rainwater used for plant growth and 'blue water use' for the use of ground- and surface water for irrigation. Both green and blue water use can be quantified in terms of volumes used per year. The impact on water quality is quantified here and made comparable to the impacts of water use by translating the volumes of emitted chemicals into the dilution volume necessary to assimilate the pollution. In the industrial stage, there are two major impacts on water: abstraction of process water from surface or groundwater (blue water use), and pollution of water as a result of the waste flows from the cotton processing industries. The latter is again translated into a certain volume of dilution water requirement.

Virtual water content of cotton products

The virtual water content of seed cotton

In order to assess the water footprint of cotton consumption in a country it is necessary to know the use of domestic water resources for domestic cotton growth or processing and also to know the water use associated with the import and export of raw cotton or cotton products. The total water footprint of a country includes two components: the part of the footprint that falls inside the country (internal water footprint) and the part of the footprint that presses on other countries in the world (external water footprint). The distinction refers to use of domestic water resources versus the use of foreign water resources (Chapagain and Hoekstra, 2004).

International trade of commodities brings along international flows of 'virtual water' (Hoekstra and Hung, 2005). 'Virtual water' is thereby defined as the volume of water used to produce a commodity (Allan, 1997; 1998a). 'Virtual water' has also been called 'embedded water' and is a similar concept as 'embodied energy', which has been defined as the direct and indirect energy required to produce a good, service or entity (Herendeen, 2004). In accounting virtual water flows the track of which parts of these flows refer to green, blue and dilution water is kept respectively.

The virtual water content of seed cotton (m^3/ton) has been calculated as the ratio of the volume of water (m^3/ha) used during the entire period of crop growth to the corresponding crop yield (ton/ha). The volume of water used to grow crops in the field has two components: effective rainfall (green water) and irrigation water (blue water). The CROPWAT model (FAO, 2003c) has been used to estimate the effective rainfall and the irrigation requirements per country. The climate data have been taken from FAO (2003b; 2003e) for the most appropriate climatic stations (USDA/NOAA, 2005b) located in the major cotton producing regions of each country. The actual irrigation water use is taken equal to the irrigation requirements as estimated with the CROPWAT model for those countries where the whole harvesting area is reportedly irrigated. In the countries where only a certain fraction of the harvesting area is irrigated, the actual irrigation water use is taken equal to this fraction times the irrigation water requirements.

The 'green' virtual water content of the crop has been estimated as the ratio of the effective rainfall to the crop yield (Equation 2). The 'blue' virtual water content of the crop has been taken equal to the ratio of the volume of irrigation water used to the crop yield (Equation 3). The total virtual water content of seed cotton is the sum of the green and blue components, calculated separately for the fifteen largest cotton-producing countries. These countries contribute nearly 90% of the global cotton production (Table 8.1). For the remaining countries the global average virtual water content of seed cotton has been assumed. In the fifteen largest cotton-producing countries, the major cotton-producing regions have been identified (Table 8.2) so that the appropriate climate data could be selected. For regions with more than one climate station, the data for the relevant stations have been equally weighed assuming that the stations represent equally sized cotton-producing areas. National average crop water requirements have been calculated on the basis of the respective share of each region to the national production.

The calculated national average crop water requirements for the fifteen largest cotton-producing countries are presented in Table 8.3. Total volumes of water use and the average virtual water content of seed cotton for the major cotton-producing countries are presented in Table 8.4. The global average virtual water content of seed cotton is 3644 m^3/ton. The global volume of water use for cotton crop production is 198 Gm^3/yr with nearly an equal share of green and blue water.

Table 8.1. The top-fifteen of seed cotton producing countries. Period 1997-2001.

Countries	Average production (ton/yr)*	% contribution to global production*	Planting period**	Yield (ton/ha)*
China	13,604,100	25.0	April/May	3.16
USA	9,699,662	17.8	March/May	1.86
India	5,544,380	10.2	April/May/July	0.62
Pakistan	5,159,839	9.5	May/June	1.73
Uzbekistan	3,342,380	6.1	April	2.24
Turkey	2,199,990	4.0	April/May	3.12
Australia	1,777,240	3.3	October/November	3.74
Brazil	1,613,193	3.0	October	2.06
Greece	1,253,288	2.3	April	3.02
Syria	1,016,594	1.9	April/May	3.92
Turkmenistan	954,440	1.8	March/April	1.72
Argentina	712,417	1.3	October/December	1.16
Egypt	710,259	1.3	February/April	2.39
Mali	463,043	0.9	May/July	1.03
Mexico	453,788	0.8	April	2.98
Others	5,939,363	10.9	-	-
World	54,443,977	100	-	-

* Source: FAOSTAT (2005).
** Sources: UNCTAD (2005b); FAO (2005); Cotton Australia (2005).

Table 8.2. Main regions of cotton production within the major cotton producing countries.

Country	Major cotton harvesting regions and their share to the national harvesting area*
Argentina	Chaco (85%)
Australia	Queensland (23%) and New Southwales (77%)
Brazil	Parana (43%), Sao Paulo (21%), Bahia (8%), Minas Gerais (5%), Mato Grosso (5%), Goias (4%) and Mato Gross do Sul (4%)
China	Xinjiang (21.5%), Henan (16.6%), Jiangsu (11.5%), Hubei (11.4%), Shandong (10%), Hebei (6.7%), Anhui (6.4%), Hunan (5.2%), Jiangxi (3.3%), Sichuan (2.3%), Shanxi (1.7%), and Zhejiang (1.3%)
Egypt	Cairo (85%)
Greece	C. Macedonia (14%), E. Macedonia (27%), and Thessaly (51%)
India	Punjab (18%), Andhra Pradesh (14%), Gujarat (14%), Maharastha (13%), Haryana (10%), Madhya Pradesh (10%), Rajasthan (8%), Karnataka (8%), and Tamil Nadu (4%)
Mali	Segou (85%)
Mexico	Baja California, Chihuahua and Coahuila
Pakistan	Sindh (15%) and Punjab (85%)
Syria	Al Hasakah (33%), Ar Raqqah (33%) and Dayr az Zawr (33%)
Turkey	Aegean/Izmir (33.6%), Antalya (1.2%), Cukurova (20.2%) and Southeasten Anotolia (45%)
Turkmenistan	Ahal (85%)
USA	North Carolina (5.4%), Missouri, Mississippi, W. Tennessee, E. Arkansas, Louisiana, Georgia (Macon) (27.7%), Georgia (Macon) (9.6%), E. Texas (33.7%) and California, Arizona (14.3%)
Uzbekistan	Fergana (85%)

* Source: USDA/NOAA (2005a).

Table 8.3. Consumptive water use at field level for cotton production in the major cotton producing countries.

	Crop water requirement (mm)	Effective rainfall (mm)	Blue water requirement (mm)	Irrigated share of area * (%)	Consumptive water use		
					Blue water (mm)	Green water (mm)	Total (mm)
Argentina	877	615	263	100	263	615	877
Australia	901	322	579	90	521	322	843
Brazil	606	542	65	15	10	542	551
China	718	397	320	75	240	397	638
Egypt	1009	0	1009	100	1009	0	1009
Greece	707	160	547	100	547	160	707
India	810	405	405	33	134	405	538
Mali	993	387	606	25	151	387	538
Mexico	771	253	518	95	492	253	746
Pakistan	850	182	668	100	668	182	850
Syria	1309	34	1275	100	1275	34	1309
Turkey	963	90	874	100	874	90	963
Turkmenistan	1025	69	956	100	956	69	1025
USA	516	311	205	52	107	311	419
Uzbekistan	999	19	981	100	981	19	999

* Sources: Gillham *et al.* (1995); FAO (1999); Cotton Australia (2005); CCI (2005); WWF (1999).

Table 8.4. Volume of water use and virtual water content of seed cotton. Period: 1997-2001.

	Volume of water use (Gm^3/yr)			Seed cotton production (ton/yr)	Virtual water content (m^3/ton)		
	Blue	Green	Total		Blue	Green	Total
Argentina	1.6	3.8	5.5	712,417	2,307	5,394	7,700
Australia	2.5	1.5	4	1,777,240	1,408	870	2,278
Brazil	0.1	4.2	4.2	1,613,193	46	2,575	2,621
China	10.3	17.1	27.5	13,604,100	760	1,258	2,018
Egypt	3	0	3	710,259	4,231	0	4,231
Greece	2.3	0.7	2.9	1,253,288	1,808	530	2,338
India	11.9	36.1	48	5,544,380	2,150	6,512	8,662
Mali	0.7	1.7	2.4	463,043	1,468	3,750	5,218
Mexico	0.8	0.4	1.1	453,788	1,655	852	2,508
Pakistan	19.9	5.4	25.4	5,159,839	3,860	1,054	4,914
Syria	3.3	0.1	3.4	1,016,594	3,252	88	3,339
Turkey	6.2	0.6	6.8	2,199,990	2,812	288	3,100
Turkmenistan	5.3	0.4	5.7	954,440	5,602	407	6,010
USA	5.6	16.2	21.8	9,699,662	576	1,673	2,249
Uzbekistan	14.6	0.3	14.9	3,342,380	4,377	83	4,460
Sub-total	88.2	88.6	176.8	48,504,613	-	-	-
Average	-	-	-	-	1,818	1,827	3,644
Other countries	10.8	10.8	21.6	5,939,363	-	-	-
World	99.0	99.4	198.4	54,443,977	-	-	-

The water use for cotton production differs considerably over the countries. Climatic conditions for cotton production are least attractive in Syria, Egypt, Turkmenistan, Uzbekistan and Turkey because evaporative demand in all these countries is very high (1000-1300 mm) while effective rainfall is very low (0-100 mm). The shortage of rain in these countries has been solved by irrigating the full harvesting area.

Resulting yields vary from world-average (Turkmenistan) to very high (Syria, Turkey). Climatic conditions for cotton production are most attractive in the USA and Brazil. Evaporative demand is low (500-600 mm), so that vast areas can suffice without irrigation. Yields are a bit above world-average. India and Mali take a particular position by producing cotton under high evaporative water demand (800-1000 mm), short-falling effective rainfall (400 mm), and partial irrigation only (between a quarter and a third of the harvesting area), resulting in relatively low overall yields.

The average virtual water content of seed cotton in the various countries gives a first rough indication of the relative impacts of the various production systems on water. Cotton from India, Argentina, Turkmenistan, Mali, Pakistan, Uzbekistan, and Egypt is most water-intensive. Cotton from China and the USA on the other hand is very water-extensive. Since blue water generally has a much larger opportunity cost than green water, it makes sense to particularly look at the blue virtual water content of cotton in the various countries. China and the USA then still show a positive picture in this comparative analysis. Also Brazil comes in a positive light now, due to the acceptable yields under largely rain-fed conditions. The blue virtual water content and thus the impact per unit of cotton production are highest in Turkmenistan, Uzbekistan, Egypt, and Pakistan, followed by Syria, Turkey, Argentina and India.

It is interesting to compare neighbouring countries such as Brazil-Argentina and India-Pakistan. Cotton from Brazil is preferable over cotton from Argentina from a water resources point of view because growth conditions are better in Brazil (smaller irrigation requirements) and even despite the fact that the cotton harvesting area in Argentina is fully irrigated (compared to 15% in Brazil), the yields in Argentina are only half the yield in Brazil. Similarly, cotton from India is to be preferred over cotton from Pakistan – again from a water resources point of view only – because the effective rainfall in Pakistan's cotton harvesting area is low compared to that in India and the harvesting area in Pakistan is fully irrigated. Although India achieves very low cotton yields per hectare, the blue water requirements per ton of product are much lower in India compared to Pakistan.

The virtual water content of cotton products
The different processing steps that transform the cotton plant through various intermediate products to some final products are shown in Figure 8.1. The virtual water content of seed cotton is attributed to its products following the methodology as introduced and applied by Chapagain and Hoekstra (2004). That means that the virtual water content of each processed cotton product has been calculated based on the product fraction (ton of crop product obtained per ton of primary crop, Equation 35) and the value fraction (the market value of the crop product divided by the aggregated market value of all crop products derived from one primary crop, Equation 36). The product fractions have been taken from the commodity trees in FAO (2003f) and UNCTAD (2005a). The value fractions have been calculated based on the market prices of the various products. The global average market prices of the cotton products have been calculated from ITC (2004). In calculating the virtual water content of fabric, the process water requirements for bleaching, dying and printing have been added (30 m^3 per ton for bleaching, 140 m^3 per ton for dying and 190 m^3 per ton for printing). In the step of finishing there is also additional water required (140 m^3 per ton). The process water requirements have to be

understood as rough average estimates, because the actual water requirements vary considerably among various techniques used (Ren, 2000).

The green and blue virtual water content of different cotton products for the major cotton producing countries is presented in Table 8.5. These water volumes do not yet include the volume of water necessary to dilute the fertiliser-enriched return flows from the cotton plantations and the polluted return flows from the processing industries.

Table 8.5. Virtual water content of cotton products at different stages of production for the major cotton producing countries (m^3/ton).

	Cotton lint		Grey fabric		Fabric		Final textile		
	Blue	Green	Blue	Green	Blue	Green	Blue	Green	Total
Argentina	5,385	12,589	5,611	13,118	5,971	13,118	6,107	13,118	19225
Australia	3,287	2,031	3,425	2,116	3,785	2,116	3,921	2,116	6037
Brazil	107	6,010	112	6,263	472	6,263	608	6,263	6870
China	1,775	2,935	1,849	3,059	2,209	3,059	2,345	3,059	5404
Egypt	9,876	0	10,291	0	10,651	0	10,787	0	10787
Greece	4,221	1,237	4,398	1,289	4,758	1,289	4,894	1,289	6183
India	5,019	15,198	5,230	15,837	5,590	15,837	5,726	15,837	21563
Mali	3,427	8,752	3,571	9,120	3,931	9,120	4,067	9,120	13188
Mexico	3,863	1,990	4,026	2,073	4,386	2,073	4,522	2,073	6595
Pakistan	9,009	2,460	9,388	2,563	9,748	2,563	9,884	2,563	12447
Syria	7,590	204	7,909	213	8,269	213	8,405	213	8618
Turkey	6,564	672	6,840	701	7,200	701	7,336	701	8037
Turkmenistan	13,077	951	13,626	991	13,986	991	14,122	991	15112
USA	1,345	3,906	1,401	4,070	1,761	4,070	1,897	4,070	5967
Uzbekistan	10,215	195	10,644	203	11,004	203	11,140	203	11343
Global average	4,242	4,264	4,421	4,443	4,781	4,443	4,917	4,443	9359

Impact on the water quality in the cotton producing countries

Impact due to use of fertilisers in crop production

Cotton production affects water quality both in the stage of growing and the stage of processing. The impact in the first stage depends upon the amount of fertilizers used and the plant fertilizer uptake rate. The latter depends on the soil type, available quantity of fertilizer and stage of plant growth. The total quantity of pesticides used, in almost all cases, gets into either ground water or surface water bodies. Only 2.4 percent of the world's arable land is planted with cotton yet cotton accounts for 24 percent of the world's insecticide market and 11 percent of the sale of global pesticides (WWF, 2003). The nutrients (nitrogen, phosphorus, potash and other minor nutrients) and pesticides that leach out of the plant root zone can contaminate groundwater and surface water.

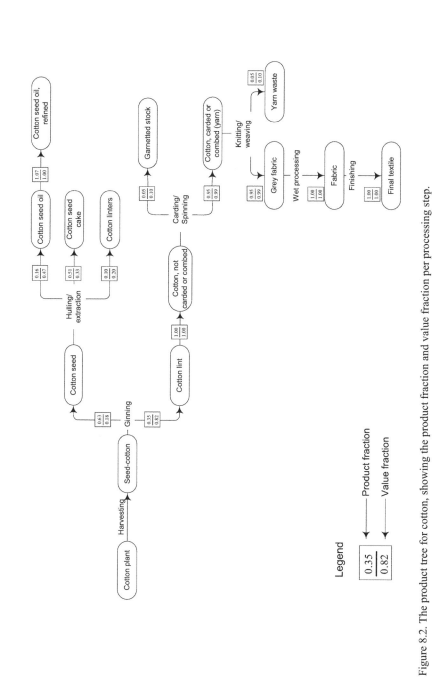

Figure 8.2. The product tree for cotton, showing the product fraction and value fraction per processing step.

The nitrite ions (NO_2-) in blood can inactivate haemoglobin, reducing the oxygen carrying capacity of the blood and the infants under 3 months are at risk. Nitrates in the drinking water can be harmful as the nitrite ions are formed in the gastrointestinal tract by the chemical reduction of the nitrate ions. Hence the target of the regulation is the nitrate intake. In surface waters, fertilizers can stimulate growth of algae and other aquatic plants, which results in a reduction of dissolved oxygen in the water when dead plant material decomposes (a process known as eutrophication).

Phosphorus has low mobility in the soil and leaching is generally not a problem. Phosphates can react with other minerals in the soil forming insoluble compounds and the amount of potassium leached is influenced by the cation exchange capacity of the soil. Instead, mobility to the roots is the prime limitation to uptake. Potassium mobility in soils is intermediate between nitrogen and phosphorus, but is not easily leached because it has a positive charge ($K+$) which causes it to be attracted to negatively charged soil colloids.

The main nitrogen processes in the soil are immobilisation/mineralization from organic matter, adsorption/desorption form cation-anion exchange sites on clay and organic matter and the application from external sources. The nitrogen is lost in various forms such as seed cotton, de-nitrification, leaching, volatilisation and burning stubble. Nitrogen is most susceptible to leaching because it cannot be retained by the soil. The nitrate ion, NO_3^- is not strongly held to clay and organic matter and is subject to movement within the soil profile. Downward movement of ions (leaching) is a problem in coarse-textured soils (loams and sands). In clay soils where movement of soil water is slow, nitrate movement is also slow. Greater losses occur from poorly structured or poorly drained soils compared to well-structured and well drained soils. The loss of fertilizer N during crop growth is variable and site dependent. Deep drainage and nutrient leaching are significant under irrigated cotton. During flood irrigation, surface soil high in nitrate is washed into cracks with the irrigation.

About 60 percent of the total nitrogen applied is removed in the seed cotton (CRC, 2004). Silvertooth *et al.* (2001) approximated that out of the total nitrogen applied to 80 percent of it gets recovered in the cotton field. The residual fraction either goes to the atmosphere by de-nitrification or discharges to the free flowing water bodies. In the present study, the quantity of N that reaches free flowing water bodies is assumed to be 10 percent of the applied rate assuming a steady state balance at root zone in the long run. The effect of use of pesticides and herbicides in cotton farming to the environment has not been analysed.

The total volume of water required per ton N is calculated considering the volume of nitrogen leached (ton/ton) and the permissible limit (ton/m^3) in the free flowing surface water bodies. The standard recommended by EPA (2005) for nitrate in drinking water is 10 milligrams per litre (measured as nitrogen) and has been taken to calculate the necessary dilution water volume. This is a conservative approach, since natural background concentration of N in the water used for dilution has been assumed negligible.

In this study, the average rate of fertiliser application for the year 1998 has been taken as reported by IFA *et al.* (2002). The total volume of fertilizer applied is calculated based on the average area of cotton harvesting for the concerned period (Table 8.6).

Table 8.6. Fertilizer application and the volume of water required to dilute the fertilizers leached to the water bodies. Period: 1997-2001.

Countries	Average fertilizer application rate* (kg/ha)			Total fertilizer applied (ton/yr)			Nitrogen leached to the water bodies	Volume of dilution water required	
	N	P₂O₅	K₂0	N	P₂O₅	K₂0	(ton/yr)	(10⁶ m³/yr)	(m³/ton)
Argentina	40	5		25,009	3,126		2,501	157	351
Australia	121	20	12.4	58,087	9,601	5,953	5,809	581	327
Brazil	40	50	50	30,674	38,342	38,342	3,067	307	190
China	120	70	25	516,637	301,372	107,633	51,664	5,166	380
Egypt	54	57	57	16,076	16,969	16,969	1,608	1,175	226
Greece	127	39	3.5	52,630	16,162	1,450	5,263	526	420
India	66	28	6	588,675	249,741	53,516	58,868	5,887	1,062
Mali	35			15,710			1,571	161	339
Mexico	120	30		18,315	4,579		1,831	183	404
Pakistan	180	28	0.4	536,720	83,490	1,193	53,672	5,367	1,040
Syria	50	50		12,964	12,964		1,296	130	128
Turkey	127	39	3.5	89,927	27,615	2,478	8,993	899	409
Turkmenistan	210	45	1.2	117,495	25,178	671	11,750	250	1,231
USA	120	60	85	625,544	312,772	443,094	62,554	6,255	645
Uzbekistan	210	45	1.2	313,274	67,130	1,790	31,327	3,133	937
Average**	91	35	20						622
Sum				3,017,737	1,169,041	673,090	301,774	30,177	

* Source: IFA *et al.* (2002). For Uzbekistan, Mali and Turkey, the fertiliser application rate has been taken from Turkmenistan, Nigeria and Greece respectively.
**The global average fertilizer application rate has been calculated from the country-specific rates, weighted on the basis of the share of a country in the global area of cotton production.

Impact due to use of chemicals in the processing stage
The average volumes of water use in wet processing (bleaching, dying and printing) and finishing stage are 360 m³/ton and 136 m³/ton of cotton textile respectively (USEPA, 1996). The biological oxygen demand (BOD), chemical oxygen demand (COD), total suspended solids (TSS) and the total dissolved solids (TDS) in the effluent from a typical textile industry are given by UNEP IE (1996) and presented in Table 8.7. In this study, the maximum permissible limits for effluents to discharge into surface and ground water bodies are taken from the guidelines set by the World Bank (1999).

As the maximum limits for different pollutants are different, the volume of water required to meet the desired level of dilution will be different per pollutant category in each production stage. Per production stage, the pollutant category that requires most dilution water has been taken as indicative for the total dilution water requirement (Table 8.8). The virtual water content of a few specific consumer products is shown in Table 8.9.

Table 8.7. Waste water characteristics at different stages of processing cotton textiles and permissible limits to discharge into water bodies.

Process	Waste water volume* (m³/ton)	Pollutants** (kg per ton of textile product)			
		BOD	COD	TSS	TDS
Wet processing	360	32	123	25	243
Bleaching	*30*	*5*	*13*		*28*
Dying	*142*	*6*	*24*		*180*
Printing	*188*	*21*	*86*	*25*	*35*
Finishing	136	6	25	12	17
Total	496	38	148	37	260
Permissible limits (milligrams per litre)***		50	250	50	

* Source: USEPA (1996)
** Source: UNEP IE (1996)
*** Source: WB (1999)

Table 8.8. Volume of water necessary to dilute pollution per production stage.

Stage of production	Volume of water per pollutant category (m³/ton of cotton textile)			Dilution water volume (applicable) (m³/ton)
	BOD	COD	TSS	
Wet processing	640	492	500	640
Finishing	120	100	240	240
Wet processing and finishing carried at the same place	760	592	740	760
Wet processing and finishing carried at different place	-	-	-	880

Table 8.9. Global average virtual water content of some selected consumer products.

	Standard weight (gram)	Virtual water content (litres)			
		Blue water	Green water	Dilution water	Total volume of water
1 pair of Jeans	1,000	4,900	4,450	1,500	10,850
1 Single bed sheets	900	4,400	4,000	1,350	9,750
1 T-shirt	250	1,230	1,110	380	2,720
1 Diaper	75	370	330	110	810
1 Johnson's cotton bud	0.333	1.6	1.5	0.5	3.6

International virtual water flows

Virtual water flows between nations have been calculated by multiplying commodity trade flows by their associated total virtual water content (Equation 40). We have taken into account the international trade of cotton products for the complete set of countries from the Personal Computer Trade Analysis System of the International Trade Centre, produced in collaboration with UNCTAD/WTO. It covers trade data from 146 reporting countries disaggregated by product and partner countries for the period 1997-2001 (ITC, 2004).

For the calculation of international virtual water flows, all cotton products are considered as reported in the database of ITC (2004). It includes the complete set of cotton products from the commodity groups 12, 14, 15, 23, 60, 61, 62 and 63. From group 52, only those products with more than 85 percent of cotton in their composition are considered.

The calculated virtual water flows between countries in relation to the international trade in cotton products add up to 204 Gm^3/yr at a global scale (an average for the period 1997-2001). About 43% of this total flow refers to blue water, about 40% to green water and about 17% to dilution water (Tables 8.10 and 8.11). The virtual water flows in relation to international trade in all crop, livestock and industrial products add up to 1625 Gm^3/yr at a global scale (Chapagain and Hoekstra, 2004). The global sum of annual gross virtual water flows between nations related to cotton trade is thus 12 per cent of the total sum of international virtual water flows.

The countries producing more than 90 percent of seed cotton are responsible for only 62 percent of the global virtual water exports (Table 8.10). This can be understood from the fact that the countries that import the raw cotton from the major producing countries export significant volumes again to other countries, often in some processed form. Export of cotton products made from imported raw cotton is significant for instance in Japan, the European Union, and Canada.

Table 8.10. Gross virtual water export from the major cotton producing countries related to export of cotton products. Period: 1997-2001.

	Green water (Gm^3/yr)	Blue water (Gm^3/yr)	Dilution water (Gm^3/yr)	Total (Gm^3/yr)	Contribution to the global flows
Argentina	1.98	0.85	0.13	2.95	1%
Australia	1.44	2.34	0.55	4.34	2%
Brazil	1.03	0.07	0.17	1.27	1%
China	11.36	9.32	5.43	26.11	13%
Egypt	-	1.72	0.13	1.85	1%
Greece	0.41	1.41	0.36	2.18	1%
India	16.83	5.75	3.08	25.66	13%
Mali	1.17	0.46	0.11	1.73	1%
Mexico	1.04	2.23	0.86	4.13	2%
Pakistan	2.87	10.64	3.05	16.56	8%
Syria	0.04	1.63	0.07	1.75	1%
Turkey	0.40	4.08	0.89	5.37	3%
Turkmenistan	0.10	1.41	0.31	1.83	1%
Uzbekistan	0.15	7.74	1.66	9.55	5%
USA	11.18	4.34	5.18	20.70	10%
Others	31.06	32.73	13.83	77.62	38%
Global flows	81.05	86.72	35.83	203.6	

Pakistan, China, Uzbekistan and India are the largest exporters of blue water. These countries export a lot of water in absolute sense, but in relative sense as well: more than half of the blue water used for cotton irrigation enters export products. The USA also appears in the top-list of total virtual water exporters due to its large share of green water export. The largest gross dilution volume exporters are China, USA and Pakistan, implying that the international trade in cotton products are having larger impact on the water quality in these countries.

Table 8.11. Largest gross virtual water importers (Gm³/yr) related to the international trade of cotton products. Period: 1997-2001.

	Green water (Gm³/yr)	Blue water (Gm³/yr)	Dilution water (Gm³/yr)	Total (Gm³/yr)	Contribution to the global flows
Brazil	2	1.5	0.4	3.9	2%
Canada	1.6	1	0.6	3.2	2%
China	15.6	15.9	6.7	38.2	19%
France	2.4	3.2	1.2	6.8	3%
Germany	3.5	5	1.8	10.4	5%
Indonesia	1.9	2	0.7	4.6	2%
Italy	2.9	4.5	1.3	8.7	4%
Japan	3.3	3.3	1.5	8.2	4%
Korea Rep.	2.6	2.8	1	6.4	3%
Mexico	6.4	2.9	3.2	12.5	6%
Netherlands	1.4	1.6	0.7	3.7	2%
Russian federation	0.5	2.5	0.6	3.7	2%
Thailand	1.5	1.4	0.5	3.3	2%
Turkey	1.4	2.6	0.7	4.7	2%
UK	2.9	3.1	1.3	7.3	4%
USA	10	12.2	5.3	27.5	14%
Others	21.2	21.1	8.3	50.6	25%
Global flows	81.05	86.72	35.83	203.6	

Water footprints related to consumption of cotton product

In assessing a national water footprint due to domestic cotton consumption, distinction has been made between the internal and the external footprint. The internal water footprint is defined as the use of domestic water resources to produce cotton products consumed by inhabitants of the country. It is the sum of the total volume of water used from the domestic water resources to produce cotton products minus the total volume of virtual water export related to export of domestically produced cotton products. The external water footprint of a country is defined as the annual volume of water resources used in other countries to produce cotton products consumed by the inhabitants of the country concerned. The external water footprint is calculated by taking the total virtual water import into the country and subtracting the volume of virtual water exported to other countries as a result of re-export of imported products.

The global water footprint related to the consumption of cotton products is estimated at 256 Gm³/yr, which is 43 m³/yr per capita in average. About 42% of this footprint is due to the use of blue water, another 39% to the use of green water and about 19% to the dilution water requirements (Table 8.12). About 44% of the global water use for cotton growth and processing is not for serving the domestic market but for export. If we do not consider the water requirements for cotton products only, but take into account the water needs for the full scope of consumed goods and services, the global water footprint is 7450×10^9 m³/yr (Chapagain and Hoekstra, 2004). This includes the use of green and blue water for the full spectrum of the global consumption goods and services, but it excludes the water requirement for dilution of waste flows. As a proxy for the latter we take here the rough estimate provided by Postel et al. (1996), who estimate the global dilution water requirement at 2350×10^9 m³/yr. This means that the full global water footprint is about 9800×10^9 m³/yr. The global water footprint related to cotton consumption is 256×10^9 m³/yr, which means that the consumption of cotton products takes a share of 2.6 per cent of the full global water footprint.

Table 8.12. The global water footprint due to cotton consumption (Gm^3/yr). Period: 1997-2001.

	Blue water footprint	Green water footprint	Dilution water footprint	Total water footprint	Contribution to the total water footprint
Internal water footprint*	59.6	54.8	28.5	143	56 %
External water footprint*	48.0	44.7	20.7	113	44 %
Total water footprint	108	99	49	256	
Contribution to the total water footprint	42 %	39 %	19 %		

* The internal water footprint at global scale refers to the aggregated internal water footprints of all nations of the world. The external water footprint refers here to the aggregated external water footprints of all nations.

The countries with the largest impact on the foreign water resources are China, USA, Mexico, Germany, UK, France, and Japan. About half of China's water footprint due to cotton consumption is within China (the internal water footprint); the other half (the external footprint) presses in other countries, mainly in India (dominantly green water use) and Pakistan (dominantly blue water use).

Per country, the water footprint as a result of domestic cotton consumption can be mapped as has been done for the USA in Figure 8.3. The arrows show the tele-connections between the area of consumption (the USA) and the areas of impact (notably India, Pakistan, China, Mexico and Dominican Republic). The total water footprint of an average US citizen due to the consumption of cotton products is 135 m^3/yr – more than three times the global average – out of which about half is from the use of external water resources. If all world citizens would consume cotton products at the US rate, other factors remaining equal, the global water use would increase by five per cent (from 9800 to 10300 Gm^3/yr), which is quite substantial given that humanity already uses more than half of the runoff water that is reasonably accessible (Postel et al., 1996).

For proper understanding of the impact map shown in Figure 8.3, it should be observed here that the map shows the full internal water footprint of the USA plus the external water footprints in other countries insofar easily traceable. For instance, USA imports several types of cotton products from the EU, that together contain 430 million m^3/yr of virtual water, but these cotton products do not fully originate from the EU25. In fact, the EU25 imports raw cotton, grey fabrics and final products from countries such as India, Uzbekistan and Pakistan, then partly or fully processes these products into final products and ultimately exports to the USA. Out of the 430 million m^3/yr of virtual water exported from the EU25 to the USA, only 16% is actually water appropriated within the EU25; the other 84% refers to water use in countries from which the EU25 imports (e.g. India, Uzbekistan, Pakistan). For simplicity, only the 'direct' external footprints (tracing the origin of imported products only one step back) are shown in the map, and not the 'indirect' external footprints. Adding the latter would mean adding for instance an arrow from India to EU25, which then is forwarded to the USA. Doing so for all indirect external water footprints would create an incomprehensible map. For the same reason, only arrows for the largest virtual water flows towards the USA are shown.

The water footprint as a result of cotton consumption in Japan is mapped in Figure 8.4. For their cotton the Japanese consumers most importantly rely on the water resources of China, Pakistan, India, Australia and the USA. Japan does not grow cotton, and also does not have a large cotton processing industry. The Japanese water footprint due to consumption of cotton products is 4.6 Gm^3/yr, of which 95 percent presses in other countries. The cotton products imported from Pakistan put a large pressure on Pakistan's scarce blue water resources. In China and even more so in India, cotton is produced with lower inputs of blue water (in relation to the green water inputs), so that cotton products from China and India put less stress per unit of cotton product on the scarce blue water resources than in Pakistan.

Figure 8.5 shows the water footprint due to cotton consumption in the twenty-five countries of the European Union (EU25). 84% of EU's cotton-related water footprint lies outside the EU. From the map it can be seen that, for their cotton supply, the European community most heavily depends on the water resources of India. This puts stress on the water availability for other purposes in India. In India one third of the cotton harvest area is being irrigated; particularly cotton imports from these irrigated areas have a large opportunity cost, because the competition for blue water resources is higher than for the green water resources. If we look at the impacts of European cotton consumption on blue water resources, the impacts are even higher in Uzbekistan than in India. Uzbekistan uses 14.6 Gm^3/yr of blue water to irrigate cotton fields, out of which it exports 3.0 Gm^3/yr in virtual form to the EU25. The consumers in the EU25 countries thus indirectly (and mostly unconsciously) contribute for about 20 per cent to the desiccation of the Aral Sea. In terms of pollution, cotton consumption in the EU25 has largest impacts in India, Uzbekistan, Pakistan, Turkey and China. These impacts are partly due to the use of fertiliser in the cotton fields and partly to the use of chemicals in the cotton processing industries. Cotton consumption in the EU25 also causes pollution in the region itself, mainly from the processing of imported raw cotton or grey fabrics into final products.

The three components of a water footprint – green water use, blue water use and dilution water requirement – affect water systems in different ways. Use of blue water generally affects the environment more than green water use. Blue water is lost to the atmosphere where otherwise it would have stayed in the ground or river system where it was taken from. Green water on the other hand would have been evaporated through another crop or through natural vegetation if it would not have been used for cotton growth. Therefore there should generally be more concern with the 'blue water footprint' than with the 'green water footprint'. The part of the water footprint that refers to dilution water requirements deserves attention as well, since pollution is a choice and not necessary. Waste flows from cotton industries can be treated so that no dilution water would be required at all. An alternative to treatment of waste flows is reduction of waste flows. With cleaner production technology, the use of chemicals in cotton industries can be reduced by 30 per cent, with a reduction of the COD content in the effluent of 60 percent (Visvanathan et al., 2000).

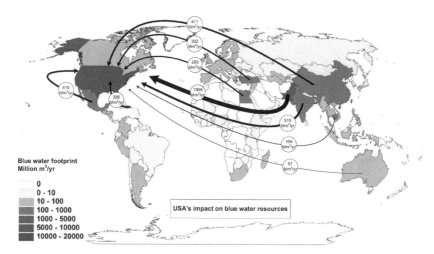

Figure 8.3a. The impact of consumption of cotton products by US citizens on the world's blue water resources (Mm³/yr). Period: 1997-2001.

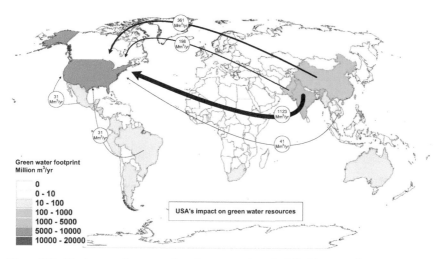

Figure 8.3b. The impact of consumption of cotton products by US citizens on the world's green water resources (Mm³/yr). Period: 1997-2001.

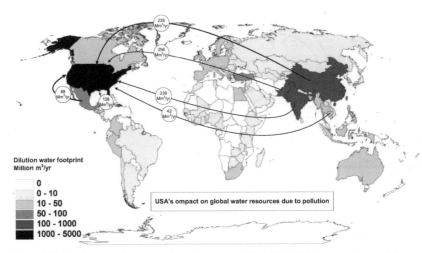

Figure 8.3c. The impact of consumption of cotton products by US citizens on the world's water resources due to pollution (Mm³/yr). Period: 1997-2001.

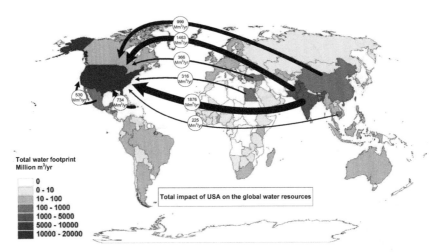

Figure 8.3d. Total impact of consumption of cotton products by US citizens on the world's water resources (Mm³/yr). Period: 1997-2001.

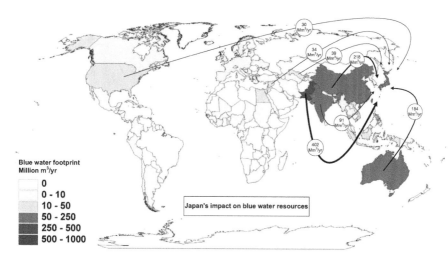

Figure 8.4a. The impact of consumption of cotton products by Japanese citizens on the world's blue water resources (Mm³/yr). Period: 1997-2001.

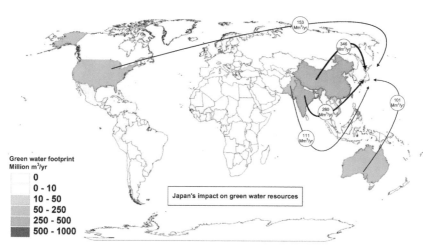

Figure 8.4b. The impact of consumption of cotton products by Japanese citizens on the world's water resources (Mm³/yr). Period: 1997-2001.

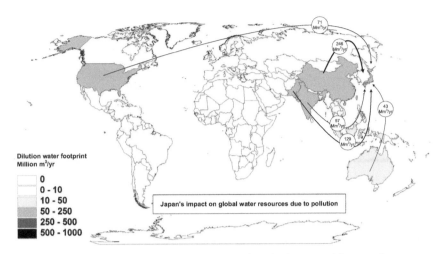

Figure 8.4c. The impact of consumption of cotton products by Japanese citizens on the world's water resources (Mm³/yr). Period: 1997-2001.

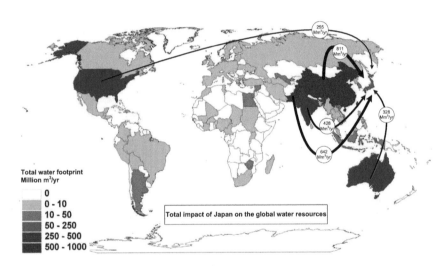

Figure 8.4d. The impact of consumption of cotton products by Japanese citizens on the world's water resources (Mm³/yr). Period: 1997-2001.

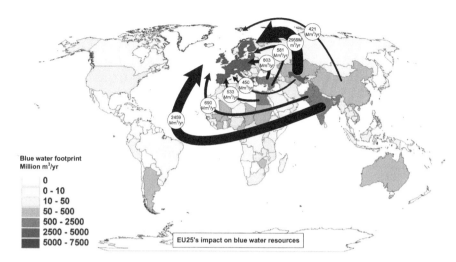

Figure 8.5a. The impact of consumption of cotton products by the people in EU25 on the world's blue water resources (Mm³/yr). Period: 1997-2001.

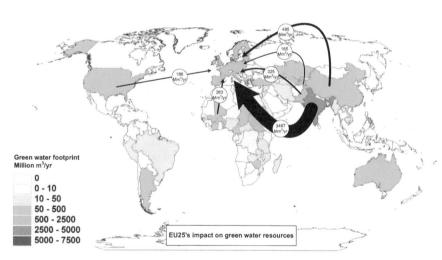

Figure 8.5b. The impact of consumption of cotton products by the people in EU25 on the world's green water resources (Mm³/yr). Period: 1997-2001.

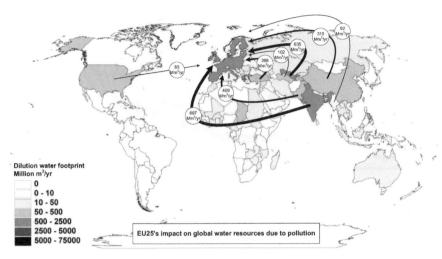

Figure 8.5c. The impact of consumption of cotton products by the people in EU25 on the world's water resources due to pollution (Mm³/yr). Period: 1997-2001.

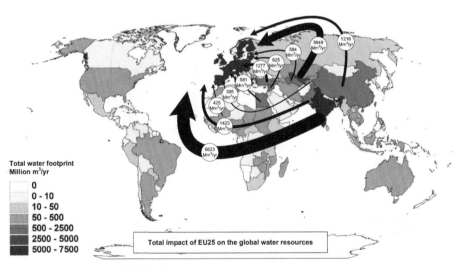

Figure 8.5d. Total impact of consumption of cotton products by the people in EU25 on the world's water resources (Mm³/yr). Period: 1997-2001.

Conclusion

There is no single indicator of sustainability, because of the variety of facts, values and uncertainties that play a role in any debate of sustainable development. The water footprint of a nation should clearly not be seen as the ultimate indicator of sustainability, but rather as a new indicator that can add to the sustainability debate. It adds to the ecological footprint and the embodied energy concept by taking water as a central viewpoint as alternative to land or energy. It adds to earlier indicators of water use by taking the consumer's perspective on water use instead of the producer's perspective.

After the introduction of the ecological footprint concept in the 1990s, several scholars have expressed doubts whether the concept is useful in science or policy making. At the same time we see that the concept attracts attention and evokes scientific debate. It is expected that the water footprint concept leads to a similar dual response. On the one hand the water footprint does not do else than gathering and presenting known data in a new format and as such does not add new knowledge. On the other hand, the water footprint adds a new fruitful perspective on issues such as water scarcity, water dependency, sustainable water use, and the implications of global trade for water management.

For water managers, water management is a river basin or catchment issue (see for instance the new South African National Water Act, 1998, and the new European Water Framework Directive, 2000). The water footprint, showing the use of water in foreign countries, shows that it is not sufficient to stick to that scale. Water problems in the major cotton producing areas of the world cannot be solved without addressing the global issue that consumers are not being held responsible for some of the economic costs and ecological impacts, which remain in the producing areas. The water footprint shows water use from the consumer's perspective, while traditional statistics show water use from the producer's perspective. This makes it possible to compare the water demand for North American or European citizens with the water demand for people in Africa, India or China. In the context of equitability and sustainability, this is a more useful comparison than a comparison between the actual water use in the USA or Europe with the actual water use in an African or Asian country, simply because the actual water use tells something about production but not about consumption.

The water footprint shows how dependent many nations are on the water resources in other countries. For its consumption of cotton products, the EU25 is very much dependent on the water resources in other continents, particularly water in Asia as this study shows, but also for other products there is a strong dependence on water resources outside Europe (Chapagain and Hoekstra, 2004). This means that water in Europe is scarcer than current indicators (showing water abstractions within Europe in relation to the available water resources within Europe) do suggest.

Cotton consumption is responsible for 2.6 per cent of the global water use. As a global average, 44 per cent of the water use for cotton growth and processing is not for serving the domestic market but for export. This means that – roughly spoken – nearly half of the water problems in the world related to cotton growth and processing can be attributed to foreign demand for cotton products. By looking at the trade relations, it is possible to track down the location of the water footprint of a community or, in other words, to link consumption at one place to the impacts at another place. The study for instance shows that the consumers in the EU25 countries indirectly contribute for about 20 per cent to the desiccation of the Aral Sea. Visualizing the actual but hidden link between cotton consumers and the water impacts of cotton production is a relevant issue in the light of the fact that the economic and environmental externalities of water use are generally not included in the price of the cotton products paid by the foreign consumers. Including information about the water footprint in product information, be it in the form of pricing or product labelling, is thus a crucial aspect in policy aimed at the reduction of negative externalities as water depletion and pollution. Given the global character of the cotton market, international cooperation in setting the rules for cotton trade is a precondition.

Since each component of the total water footprint includes a certain economic cost and environmental impact, it would be useful to see which of the costs and impacts are transferred to the consumer. In this study a careful examination of that has not been carried out, but there is quite some evidence that the majority of costs and impacts of water use and pollution caused in agriculture and industry is not translated into the price of products. According to the World Bank, the economic cost recovery in developing countries in the water sector is about 25 per cent (Serageldin, 1995). Social and environmental impacts of water use are generally not translated into the price of products at all, with sometimes an exception for the costs made for wastewater treatment before disposal. Most of the global waste flows are not treated however. Although a few industrialised countries achieve a wastewater treatment coverage of nearly 100 per cent, this coverage remains below five per cent in most developing countries (Eurostat, 2005; Hoekstra, 1998). Besides, the hundred per cent waste coverage in some of the industrialised countries refers to treatment of concentrated waste flows from households and industries only, but excludes the diffuse waste flow in agriculture. Given the general lack of proper water pricing mechanisms or other ways of transmitting production-information, cotton consumers have little incentive to take responsibility for the impacts on remote water systems.

About one fifth of the global water footprint due to cotton consumption is related to the pollution. This estimate is based on the assumption that wastewater flows can be translated into a certain water requirement for dilution based on water quality standards. Implicitly it is assumed here that the majority of waste flows enters natural water bodies without prior treatment, which is certainly true for leaching of fertilisers in agriculture and largely true for waste flows from cotton industries. In some of the rich countries, however, there is often treatment of waste flows from industries before disposal, so that we have got an overestimate of dilution water requirements here. In case of treatment of waste flows to the extent that the effluents meet water quality standards, a better estimate for the water requirement would be to consider the actual water use for the treatment process. Another issue is that the natural background concentrations in dilution water have not been accounted for, so that we have got a conservative estimate for the required dilution volume. The estimates of dilution volumes of water are also made by looking at the dilution volume required for fertilisers, but not at the volume for diluting pesticides used.

Chapter 9

Discussion

Globalisation of water: re-visiting the research questions
The research has focused on analysing the pros and cons of international trade on water resources. In particular the opportunities and threats of virtual water trade on water resources management have been explored. The main aim of the thesis is to bring out the hidden links between a consumer and a producer with respect to water demand and supply. In order to visualise the international trade of products in terms of water flows, the concept of virtual water has been used. The volume of water used in order to produce a certain amount of product has been quantified based on the local production parameters for all producing regions. The present research explicitly covers almost the entire range of agricultural products. Using the international trade data, the associated virtual water flows have been quantified. For industrial products, the virtual water flows are estimated with an aggregated approach per nation.

The role of virtual water in relieving the pressure on national water resources has been quantified along with the associated virtual water dependency. The research reflects the importance of differences in water productivities across time and space. The possibility to promote certain types of virtual water trade and to demote other types of virtual water trade in enhancing global water use efficiency has been explored; the associated risk of shifting the environmental burden from the consumption of goods has been shown along with the chain of product trade.

Different case studies have been carried out to illustrate the necessity to analyse the impact of local consumption on the global water resources, both quantitatively and qualitatively. This study further provides a set of indicators of water use from a consumer's perspective, quantifying the impact of individual or national consumption on global water resources.

Is there any global component of fresh water fulfilling the local demand?
These days the view that for studying water issues the river basin is the appropriate unit is becoming more and more common. The argument behind this thinking is that the hydrological cycle does not obey the political boundaries. Many river basins are shared between different political domains. The disturbance in one component of the hydrological cycle affects the other components. For example, upstream abstraction and consumptive use of water or pollution of water resources not only reduces the available water resources downstream but also affects the socio-economic balance of the region. This is obvious and has been the strong basis for cooperation between states sharing a common river basin. From the conventional point of view, these impacts do not seem to propagate beyond the boundaries of a river basin, assuming that the change in evaporation or runoff to the sea has very little impact on the global hydrological cycle.

A phenomenon that so far has received little attention, but that nevertheless has become relevant, is that the transfer of water intensive products across river basins creates a huge flux of virtual water connecting different river basins or nations or regions. This transfer of products not only has impact on the water resources at production sites but also at the sites where it is being imported. This research quantifies the virtual water fluxes related to the international trade of products and its impact on the local water resources. The quantification shows that 16% of the global water use is not for the domestic consumption but for export.

This cross-basin, or say international, water connections are strongest between countries with high production potentials and countries with relatively scarce water resources but rich in economic sense. The largest gross exporters are USA, Canada, France and Australia. The largest gross importers are USA, Germany, Japan and Italy. Though some largest gross exporters are the largest gross importer as well, the related products are entirely different. At regional scale, the largest exporter of virtual water is 'North America' and the largest importers are 'Western Europe' and 'Central and South Asia'. The analysis shows that in most countries fresh water demands are being met, from reasonable to significant level, not only from their own domestic water resources but from the global water resources. In this context water resources are becoming more and more a global commodity.

Can import of water in virtual form offer a relieve to water scarcity in water-poor countries?

In general, the international trade of products is indirectly relieving the pressure on water resources in water poor countries that have net import of virtual water. A good example is Jordan which is water poor but the stress is more or less relieved with the high import of water intensive products. One would assume that if a country is water-poor, it might be attractive to import virtual water, but the existing numbers do not show the relation so explicitly. This is not surprising as a country needs to have sufficient other resources, such as foreign exchange, to import products. For example, Yemen has rather *chosen* to overexploit its groundwater resources than to import goods mainly because of the absence of the possibility to import. Also, despite having little water, a country might choose to be self-sufficient to some extent in its own production system. For example, Egypt with no rainfall and completely depending on Nile water does not strongly depend on import from outside. Hence, one can conclude that besides the availability of water in a country in relation to its demand, the option to import water in its virtual form to relieve pressure on its domestic water resources is affected by a number of factors such as the availability of land, labour and other resources and the existing national policy with respect to food security.

Can export of water in virtual form contribute to problem of water scarcity?

North China experiences high water scarcity but exports 52 Gm^3/yr to South China in the form of agricultural products. The water scarcity of North China can be addressed with a wide range of possibilities including a critical analysis of the role of virtual water flows. Currently China is implementing the South-North Water Transfer Project – aiming at a real water transfer from the South to the North of 38-43 Gm^3/yr – while alternatively the water shortage in the North could be reduced also by demoting water-intensive production in the North aimed at export to South. Another case illustrating how water problems relate to virtual water export is described in the case study for cotton. The Aral Sea is drying as a result of water use from the inflowing rivers for cotton production, which is largely meant for export.

Is there any saving in global water resources as a result of trade?
Importing water intensive products saves water in the importing country that would otherwise have been consumed domestically. However at the same time, the exporting countries are losing their domestic water.

Besides national effects, the virtual water fluxes have effect on the water use efficiency globally. The research shows that at present the international trade in agricultural products alone is saving 352 billion cubic metre of water annually. This is possible as a net flow of these products are from water productive sites to the less productive sites. Since 16% of the global water use is not for the domestic consumption but for export, global water use efficiency becomes more important with increasing globalisation of trade.

The gain of one kind of water does not result in the loss of similar kind of water. For example, import of wheat in Egypt from USA or northern Europe saves blue water resources in Egypt at the cost of green water in the exporting regions. Similarly, another example is the production of crops when there is abundant water available, such as rice production during monsoon period in South Asia, which costs water in the wet period but saves water in the dry period. This acts as a real time storage dam of water resources, but in virtual form. This phenomenon of switching the kind of water saving as a result of product transfer opens the door of possibilities to reduce the gap between demand and supply of water in time and space.

How does one affect the global water resources by consuming a single product?
The coffee consumed in the Netherlands mainly comes from South America and Central America. One cup of coffee consumed in the Netherlands is equivalent of consuming 140 litres of water from those coffee exporting countries to the Netherlands. As coffee plants are primarily grown rain-fed, most of this water is green water. However, the coffee processed with the wet production method often carries significant impacts as the water used is blue water and the waste flow most often untreated.

To illustrate the importance of pollution from the production process a case study on cotton consumption has been carried out. The people in the EU25 regions get more than 84% of their cotton from outside EU25. Large part of it comes from India, Uzbekistan, and Pakistan etc. However the impacts on these countries vary both with respect to the quality and quantity. Indian cotton has a higher green to blue water use ratio than other countries. On the contrary, Uzbekistan produces cotton with 100% irrigation, and also gets less use of effective rainfall in crop production. Thus the consumption of Uzbek-cotton not only reduces the precious blue water resources from the territory, one of the basin countries of the drying Aral Sea, but also pollutes through the return flows from the agricultural fields and the processing industries. To produce one standard pair Jeans, on average it takes about 5000 litres of blue water, 4500 litres of green water and it pollutes 1500 litres from the return flows from cotton fields and cotton industries where it is processed. This number varies much depending upon the use of resources in different countries. The size of pollution is significant and it deserves attention because pollution is a choice not a compulsion. However, the choice to treat return flows then shifts to energy and other resources necessary to do so. Cotton crops in some countries (Pakistan, Syria, Turkey, Egypt, Greece and Uzbekistan) are grown in fields with hundred percent irrigation water uses and in some countries (Brazil, India, Mali and USA) partially in rain-fed fields. Hence, the impacts are different, though the number can be comparable.

The consumption of blue water does not carry negative impact all the time. For example, rice from Thailand contains both blue and green water. Though Thailand is exporting a considerable share of its available blue water resources, it is mainly from the rice production in wet season when blue water carries low opportunity cost. However, in those cases where the water for irrigation is diverted from reservoirs or from the water that is supposed to fill the reservoir or has another better alternative use, Thailand should re-visit their rice export policy relative to the benefit gained.

Why is it necessary to have a new set of indicators of water use?

To understand the impact of the global component of fresh water demand and supply, a set of indicators has been developed, such as water footprint and virtual water import dependency. Depending on the source of water use, the impacts of consumption on water resources can reside in the same country as the consumption takes place or outside this country. Accordingly, the study distinguishes two kinds of water footprints of a nation: internal water footprint and external water footprint. If all the water demands related to the production of the consumption goods in a country are met by import, the water footprint of the country is hundred percent external and also the water dependency is hundred percent. However, if all the goods are produced from the use of domestic water resources, the country has hundred percent water self sufficiency. The knowledge of the total water footprint of a country can reveal how water-scarce it is. The common methods of measuring water scarcity are production (supply) oriented; the current study shows a different perspective of water scarcity by relating the total volume of water needed to support the national demand to its available water resources.

The concept of water footprint can be used to understand a nation's real call on the natural environment. Before one can draw policy relevant information from the water footprint of a nation, detailed information on the water footprint is necessary such as which component relates to blue water use, which component to green water us and which component to water pollution. Also the opportunity cost of the water used is a relevant factor. Alternative ways to reduce the water footprint of a nation are: changing the consumption pattern from high to less water intensive products, reducing the volume of products consumed, increasing the productivity of water or importing products from places having higher water productivity. However, the effectiveness of the reduced water footprint should be seen in the context of the impacts in production sites.

Though a single indicator of sustainability does not exist, the water footprint of a nation is a new indicator that adds to the sustainability debate. It extends the idea of the ecological footprint by taking water as its central viewpoint. It is an indicator that can help to explain the effect of a consumer on the global water resources, contrary to the existing producer oriented indicators of water use.

What is the risk of becoming water dependent with increased globalisation?

Globalisation brings not only opportunities but also risks. At one hand a country can relieve the pressure on its scarce water resources by importing water intensive products instead of producing itself, but on the other hand it increases its dependency on import from other nations or regions. The importing regions must have other resources to be able to import virtual water. The high water import-dependency of many countries such as Japan (which is already 64%), is only possible because of its capacity to generate enough foreign currencies from its other resources to buy agricultural products from the international markets. This may not be a feasible option for a country which is not only poor in terms of water but also in other resources.

Becoming water dependent and having adequate purchasing power are not the only risks and challenges for water-poor countries aiming at the increase of virtual water import. One needs to analyse a few other important aspects such as: reduced employment in the agricultural sector, increased urbanisation, and reduced availability of food to the poor.

Is externalising the water footprint a threat of shifting the environmental burdens to a distant location?

The case study of cotton shows the relation between the consumption of a cotton product and the impacts at the location where it is being produced. About 44 per cent of the water used for cotton production is not for domestic consumption but to meet the export demands. Stated differently, nearly half of the water problems in the world related to the cotton production are driven by the foreign demands for cotton products. For example, the people in the EU25 region are indirectly responsible for about 20% of the desiccation of the Aral Sea. It is not argued that the cotton export from the basin has no return values. It does have economic returns in the form of gain in foreign exchange and employment opportunities in the region. However, that is not solving the water problems in the shrinking sea. As the water abstractions from the rivers that feed the Aral Sea are unsustainable, the economic benefits are also not sustainable.

One fifth of the global water use for cotton production is from pollution by return flows. The costs of this pollution are external to the polluters and as a consequence not included in the price of the cotton products. As a result, the exporting countries are left with the environmental burden that relates to the consumption in the importing regions.

Concluding remarks

International trade has been effective in improving global water use efficiency in the period 1997-2001. Given the context of increasing globalisation, it is important to prepare national or regional water policy based on an analysis of virtual water flows. In this aspect, the author believes that the appropriate scale of water resources management should always include all scales from the local to the global, unless a place is completely isolated from the rest of the world. Though understanding the water flows within the hydrological unit of a river basin is important, it provides only a partial view on the water resources context of a basin, which can be fully understood only if also the virtual flows are analysed. For sustainable use of water resources, the use of a consumption based indicator such as the 'water footprint' can be an additional tool to understand the impact of local consumption on global water resources.

The research has illustrated a new dimension in the field of managing scarce water resources. It raises a number of new and potentially interesting avenues for future research. The main areas for future explorations are: analyse existing national water policies in the light of virtual water fluxes along with the socio-economic status of a country, carry out analysis to see the importance of rain-fed agriculture in addressing the local and global water scarcity, analyse the impact of climate change and the possibilities of the virtual water storage over time to mitigate the variability in water supply, and finally carry out a water footprint analysis for the most important goods and services giving better insights into the effect of our consumption behaviour on the global water resources.

List of symbols

Symbol	Interpretation	Unit
A_c	Total harvest area of crop c in a country	ha
a_s	Angstrom constant, $=0.25$ for locations not calibrated	-
b_s	Angstrom constant, $=0.50$ for locations not calibrated	-
C	Volume of feed crop c consumed by animal a	ton/day
C_{dom}	Volume of product p consumed domestically	ton/yr
c_p	Specific heat at constant pressure, 1.013×10^{-3}	MJ/kg/°C
d_r	Inverse relative distance Earth-Sun	
D_{gross}	Gross domestic product of a nation	US$/yr
D_{ind}	Added value of the industrial sector to gross domestic product of a nation	US$/yr
e_a	Actual vapour pressure	kPa
e_s	Saturation vapour pressure	kPa
E_c	Crop evaporation in day d for crop c	m/day
E_r	Reference crop evaporation in day d	m/day
F	Green to blue water use fraction	-
f_p	Product fraction of product p	-
f_v	Value fraction of product p	-
F	Water footprint of a nation	m^3/yr
$F_{external}$	External water footprint of a country	m^3/yr
$F_{internal}$	Internal water footprint of a country	m^3/yr
F_{indv}	Water footprint of an individual	m^3/cap/yr
F_{pc}	Average per capita water footprint of a nation	m^3/cap/yr
G	Soil heat flux	MJ/m^2/day
G_{sc}	Solar constant $= 0.0820$	MJ/m^2/day
J	Number of the day in the year starting from 1 January as 1	-
K_c	Crop coefficient	-
l_p	Length of growing period of crop c	day
M_e	Export value of industrial products	US$/yr
M_i	Import value of the industrial products	US$/yr
N	Actual duration of sunshine hours	hr
N	Maximum possible duration of sunshine or daylight hours	hr
N_{pop}	Population of a country	hr
p_x	Consumption of product p by an individual	ton/yr
P	Atmospheric pressure	kPa
P_e	Effective rainfall in crop production	m^3/ha
q_{drink}	Daily drinking water consumption of animal a	m^3/day

Symbol	Interpretation	Unit
q_{mix}	Volume of water required for mixing the feed	m³/day
q_{serv}	Daily service water consumption of animal a	m³/day
Q_{proc}	Volume of processing water consumed to process crop c or animal a	m³
Q_{nat}	Renewable water resources in a country	m³/yr
R_a	Extraterrestrial radiation	MJ/m²/day
R_n	Net radiation at the crop surface	MJ/m²/day
R_{ns}	Net solar or shortwave radiation	MJ/m²/day
R_s	Incoming solar radiation	MJ/m²/day
R_{so}	Clear-sky radiation	MJ/m²/day
R_{nl}	Net outgoing longwave radiation	MJ/m²/day
R_c	Crop water requirement for the entire growth period of a crop c	m³/ha
R_{proc}	Processing water requirement per ton of primary crop c or live animal a	m³/ton
$\Delta S_{g,p,e,i}$	Global water saving resulting from export of product p from country e to country i	m³/yr
$\Delta S_{g,b}$	Global blue water saving	m³/yr
$\Delta S_{g,g}$	Global green water saving	m³/yr
$\Delta S_{n,i}$	Net national water saving of a country i as a result of trade of product p	m³/yr
T	Average air temperature	°C
T_{max}	Daily maximum temperature	°C
$T_{max,\,K}$	Maximum absolute temperature during the 24-hour period	K
T_{mean}	Daily mean air temperature	°C
T_{min}	Daily minimum temperature	°C
$T_{min,\,K}$	Minimum absolute temperature during the 24-hour period	K
$T_{month,\,i}$	Mean air temperature of month i	°C
$T_{month,\,i-1}$	Mean air temperature of previous month	°C
$T_{month,\,i+1}$	Mean air temperature of next month	°C
$T_{p,e,i}$	Volume of product exported from country e to country i	ton/yr
T_{net}	Net import of product p in country i	ton/yr
U_2	Wind speed measured at 2 m height	m/s
U	Volume of water used to produce Y unit of product	m³
U_b	Volume of blue water used for the production	m³/yr
U_c	Volume of water used for crop production	m³/yr
U_f	Volume of water used from fossil ground water for the production	m³/yr
U_{agr}	Water use in agricultural productions in a country	m³/yr
U_{dom}	Water use for household consumption in a country	m³/yr
U_{ind}	Water use in industrial products in a country	m³/yr

Symbol	Interpretation	Unit
U_{nation}	Total volume of water use in a nation	m³/yr
v_p	Market value of product p	US$/ton
V	Virtual water content of a product p	m³/unit
V_a	Virtual water content of animal a	m³/animal
V_e	Virtual water content of the product in exporting country e	m³/ton
V_i	Virtual water content of product in the importing country i	m³/ton
V_b	Blue virtual water content of a product p	m³/ton
V_c	Virtual water content of a crop c	m³/ton
$V_{c,n}$	Country average virtual water content of a crop c	m³/ton
V_f	Fossil virtual water content of a product p	m³/ton
V_g	Green virtual water content of a product p	m³/ton
$V_{a,drink}$	Virtual water content of animal a related to drinking water	m³/animal
$V_{a,feed}$	Virtual water content of animal a related to feed	m³/animal
$V_{a,serv}$	Virtual water content of animal a related to service	m³/animal
V_{ind}	Virtual water content per dollar added value in the industrial sector	m³/US$
$V_{ind,g}$	Global average virtual water content of industrial products	m³/US$
$W_{b,vir}$	Virtual water balance of a country	m³/yr
W_d	Virtual water import dependency	%
W_s	National water scarcity	%
W_{ss}	Water self-sufficiency of a country	%
Y	Volume of production	ton or US$
Y_c	Volume of production of crop c	ton/ha
z	Elevation (height above mean sea level)	m
Λ	Virtual water flow as a result of product trade	m³/yr
Λ_{ex}	Virtual water flow as a result of export of product T from country e to country i	m³/yr
Λ_{im}	Virtual water flow as a result of import of product T from country e from country i	m³/yr
Λ_{net}	Net virtual water import as a result of product trade in a country e	m³/yr
$\Lambda_{ex,dom}$	Volume of virtual water export to other countries related to export of domestically produced products	m³/yr
$\Lambda_{ind,ex}$	Virtual water export from country e related to export of industrial products	m³/yr
Λ_{re-ex}	Volume of virtual water exported to other countries as a result of re-export of imported products	m³/yr
$\Lambda_{ind,im}$	Virtual water import in country i related to import of industrial products	m³/yr
χ	Total weight of the primary crop c or live animal a processed	ton

Symbol	*Interpretation*	*Unit*
χ_{proc}	Weight of primary product p obtained from processing χ ton of primary crop c or χ ton of live animal a	ton
Δ	Slope of the vapour pressure curve	kPa/°C
γ	Psychrometric constant	kPa/°C
λ	Latent heat of vaporization, 2.45	MJ/kg
ε	Ratio molecular weight of water vapour/dry air = 0.622	-
α	Albedo or canopy reflection coefficient	-
σ	Stefan-Boltzmann constant = 4.903 x 10^{-9}	MJ/K^4/m^2/day
ω_s	Sunset hour angle	rad
φ	Latitude	rad
r_a	aerodynamic resistance	d/m
r_s	bulk surface resistance of the crop canopy and soil = 70	d/m
ρ_a	Mean air density at constant pressure	kg/m^3
ρ_w	density of water = 1000 (taken for the study)	kg/m^3
δ	Solar decimation	rad

Glossary

Blue vs. green water footprint   The total water footprint of a nation or individual falls apart into two components: the blue and the green water footprint. The blue water footprint is the volume of water withdrawn from the global blue water resources (surface water and ground water) to fulfil the national or individual demand for goods and services. The green water footprint is the volume of water used from the global green water resources (water stored in soil as soil moisture) to fulfil the demand for goods and services.

Global water saving   International trade can save water globally if a water-intensive commodity is traded from an area where it is produced with high water productivity (resulting in products with low virtual water content) to an area with lower water productivity.

Individual water footprint   The water footprint of an individual is defined as the total water used for the production of the goods and services consumed by the individual. It can be estimated by multiplying all goods and services consumed by their respective virtual water content.

Internal vs. external water footprint   The total water footprint of a country includes two components: the part of the footprint that falls inside the country (internal water footprint) and the part of the footprint that presses on other countries in the world (external water footprint). The distinction refers to the difference between the uses of domestic water resources versus the foreign water resources.

National water saving   A nation can save its domestic water resources by importing a water-intensive product rather than produce it domestically.

Nation's water footprint   The water footprint of a nation is defined as the total amount of water that is used to produce the goods and services consumed by the inhabitants of the nation. The national water footprint can be assessed in two ways. The bottom-up approach is to consider the sum of all goods and services consumed multiplied with their respective virtual water content. It should be noted that the virtual water content of a particular consumption good can vary as a function of the place and conditions of production. In the top-down approach, the water footprint of a nation can be calculated as the total use of domestic water resources plus the net virtual water import.

Virtual water content

The virtual water content of a product is the volume of water used to produce the product, measured at the place where the product was actually produced (production site specific definition). The virtual water content of a product can also be defined as the volume of water that would have been required to produce the product in the place where the product is consumed (consumption site specific definition). Unless otherwise expressed explicitly, in our studies we use the production site-specific definition. The adjective 'virtual' refers to the fact that most of the water used to produce a product is in the end not contained in the product. The real water content of products is generally negligible if compared to the virtual water content.

Virtual water balance

The virtual water balance of a country over a certain time period is defined as the net import of virtual water over this period, which is equal to the gross import of virtual water minus the gross export. A positive virtual water balance implies net inflow of virtual water to the country from other countries. A negative balance means net outflow of virtual water.

Virtual water export

The virtual water export of a country or region is the volume of virtual water associated with the export of goods or services from the country or region. It is the total volume of water required to produce the products for export.

Virtual water flow

The virtual water flow between two nations or regions is the volume of virtual water that is being transferred from one place to another as a result of product trade.

Virtual water import

The virtual water import of a country or region is the volume of virtual water associated with the import of goods or services into the country or region. It is the total volume of water required (in the export countries) to produce the products for import. Viewed from the perspective of the importing country, this water can be seen as an additional source of water that comes on top of the domestically available water resources.

Water footprint

The water footprint of an individual, business or nation is defined as the total volume of fresh water that is used to produce the goods and services consumed by the individual, business or nation. A water footprint is generally expressed in terms of the volume of water use per year.

Water import dependency	Countries with import of virtual water depend, de facto, on the water resources available in other parts of the world. The virtual water import dependency of a country or region is defined as the ratio of the external water footprint of the country or region to its total water footprint
Water scarcity	Water scarcity has often been defined as the ratio of actual water withdrawals to the available renewable water resources. This supply-oriented definition is useful from a production perspective, but does not express the scarcity from a demand perspective. In this study, water scarcity is defined as the ratio of the total water footprint of a country or region to the total renewable water resources. The national water scarcity can be more than 100% if a nation consumes more water than domestically available.
Water self-sufficiency	This is the ratio of the internal water footprint to the total water footprint of a country or region. It denotes the national capability of supplying the water needed for the production of the domestic demand for goods and services. Self-sufficiency is 100% if all the water needed is available and indeed taken from within the own territory. Water self-sufficiency approaches zero if the demand for goods and services in a country is largely met with virtual water imports.

Samenvatting

Hoewel het stroomgebied over het algemeen als aangewezen eenheid voor het analyseren van de beschikbaarheid en het gebruik van zoetwater wordt gezien, wordt het steeds belangrijker om zoetwaterkwesties in een globale context te plaatsen. De reden is dat internationale handel van producten internationale en intercontinentale overdracht van grote volumes van water in virtuele vorm met zich meebrengt. Diverse waterschaarse landen voeren opzettelijk 'virtueel water' in om de druk op de eigen binnenlandse watervoorraden te verminderen. Onder 'virtueel water' wordt hier het watervolume dat wordt gebruikt om goederen te produceren begrepen. De doelstelling van het onderzoek is het analyseren van de kansen en de bedreigingen van internationale handel in virtueel water in de context van het oplossen van nationale en regionale problemen van watertekorten. De centrale vragen waarop de studie zich richt zijn: Wat zijn de stromen van virtueel water gerelateerd aan de internationale handel van producten? Is de invoer van virtueel water een oplossing voor waterschaarse landen of juist een bedreiging, in de zin dat de landen hierdoor afhankelijk worden van waterbronnen elders? Kan internationale handel van producten een middel zijn om te komen tot een efficiënter watergebruik op mondiaal niveau, of is het een manier om de milieulasten naar een veraf gelegen locatie te verplaatsen? Om de mondiale component van de vraag naar en levering van zoetwater te begrijpen, is een reeks indicatoren ontwikkeld. Het zo ontwikkelde kader is toegepast in verschillende case studies.

De virtuele waterstromen tussen landen zijn geschat op basis van statistieken van internationale handel in producten en de virtuele waterinhoud per product in het uitvoerende land. Het berekende mondiale volume van virtuele waterstromen gerelateerd aan de internationale handel in goederen is 1625 Gm^3/jr. Ongeveer 80% van deze virtuele waterstromen heeft betrekking op de handel in landbouwproducten, terwijl de rest betrekking heeft op handel in industrieproducten. Een geschatte 16% van het mondiale watergebruik is niet voor het produceren van in eigen land geconsumeerde producten maar voor het maken van uitvoerproducten. Met een toenemende globalisering van de handel zullen de waterafhankelijkheden en de overzeese externaliteiten waarschijnlijk stijgen. Tezelfdertijd leidt de liberalisering van handel tot kansen om de mondiale watergebruikefficiëntie te verhogen en fysieke waterbesparingen te verwezenlijken.

Veel landen besparen binnenlandse waterbronnen door waterintensieve producten in te voeren en goederen uit te voeren die minder waterintensief zijn. De nationale waterbesparing door de invoer van een product kan een waterbesparing op mondiaal niveau impliceren als de stroom plaats heeft van een plaats met hoge naar een plaats met lage waterproductiviteit. Het onderzoek analyseert de gevolgen van internationale virtuele waterstromen op de mondiale en nationale waterbalansen. De studie toont aan dat de totale hoeveelheid water die in de invoerende landen vereist zou zijn als alle ingevoerde landbouwproducten in eigen land zouden zijn geproduceerd 1605 Gm^3/jr is. Deze producten worden echter geproduceerd met slechts 1253 Gm^3/jr in de uitvoerende landen, wat een mondiale waterbesparing van 352 Gm^3/jr impliceert. Deze besparing is gelijk aan 28 procent van de internationale virtuele waterstromen gerelateerd aan de handel in landbouwproducten en 6 procent van het mondiale watergebruik in de landbouw. Nationale beleidsmakers zijn echter niet geïnteresseerd in mondiale waterbesparingen maar in de toestand van de nationale watervoorraden. Egypte voert tarwe in en bespaart daardoor 3,6 Gm^3/jr van zijn nationale waterbronnen. Het gebruik van water voor het maken van exportartikelen kan voordelig zijn, zoals in Ivoorkust, Ghana en Brazilië, waar het gebruik van groen water (hoofdzakelijk in de regengevoede landbouw) voor de productie van exportgewassen zoals koffie, thee en cacao een positieve invloed heeft op de nationale economie. Het gebruik van 28 Gm^3/jr in Thailand ten behoeve van de productie van exportrijst gaat echter ten koste van extra druk op de Thaise voorraden blauw water. Het invoeren van een product met een relatief hoge groene ten opzichte van blauwe virtuele waterinhoud, bespaart mondiaal blauw water, dat over het algemeen een hogere opportunity cost heeft dan groen water.

Het gebruik van virtuele wateroverdrachten als alternatief voor echte wateroverdrachten tussen stroomgebieden is geanalyseerd in een case study voor China. Noord China heeft te kampen met ernstige waterschaarste – meer dan 40% van de jaarlijkse vernieuwbare waterhoeveelheid wordt onttrokken voor menselijk gebruik. Niettemin wordt bijna 10% van het water dat in de landbouw wordt gebruikt toegepast voor het produceren van voedsel dat naar Zuid China wordt uitgevoerd. Om deze 'virtuele waterstroom' van het noorden naar het zuiden te compenseren en de waterschaarste in het noorden te verminderen, wordt momenteel het reusachtige Zuid-Noord Water Transfer Project ten uitvoer gelegd. Deze paradox, overdracht van grote hoeveelheden water van het waterrijke zuiden naar het waterarme noorden en tezelfdertijd overdracht van grote hoeveelheden voedsel van het noorden naar het zuiden, ontvangt toenemende belangstelling, maar het onderzoek op dit gebied stagneert in het stadium van ruwe schattingen en kwalitatieve beschrijving. De huidige studie kwantificeert de volumes van virtuele waterstromen tussen de regio's in China en plaatst deze in de context van de waterbeschikbaarheid per gebied. Noord China voert jaarlijks ongeveer 52 miljard m^3 water in virtuele vorm naar Zuid China uit, wat niet veel meer is dan het maximum voorgestelde volume van de wateroverdracht langs de drie routes van het Zuid-Noord Water Transfer Project.

Om het effect van consumptie in een land op de waterbronnen in de wereld te kwantificeren en visualiseren, hanteert de studie het concept van de watervoetafdruk. De watervoetafdruk van een land is het totale volume van zoet water dat wordt gebruikt om de goederen en de diensten te maken die door de inwoners van het land worden geconsumeerd. Een watervoetafdruk wordt uitgedrukt in termen van het volume van watergebruik per jaar. De interne watervoetafdruk is het watergebruik in het betreffende land zelf, terwijl de externe watervoetafdruk het watergebruik in andere landen vertegenwoordigt. De studie heeft de watervoetafdruk voor elk land van de wereld gekwantificeerd voor de periode 1997-2001. Een Amerikaan heeft een gemiddelde watervoetafdruk van 2480 m^3 per jaar, terwijl een Chinees een gemiddelde voetafdruk van 700 m^3 per jaar heeft. De gemiddelde wereldburger heeft een watervoetafdruk van 1240 m^3 per jaar. Er zijn vier belangrijke directe factoren die de watervoetafdruk van een land bepalen: consumptievolume (gerelateerd aan het bruto nationale inkomen); consumptiepatroon (bijv. veel of weinig vlees); klimaat (de groeivoorwaarden); en landbouwpraktijk (de efficiëntie van het watergebruik).

De mondiale watervoetafdruk van koffie- en theeconsumptie is uitgewerkt met een voorbeeld voor Nederland met het onderliggende doel indicatoren te ontwikkelen die kunnen bijdragen aan het scheppen van bewustzijn over de gevolgen van consumptie voor het gebruik van natuurlijke hulpbronnen. Een standaardkop koffie of thee in Nederland kost respectievelijk ongeveer 140 liter en

34 liter water, waarvan in beide gevallen veruit het grootste deel voor het groeien van de plant. Het overgrote deel van het gebruikte water is regenwater, een bron met minder concurrentie tussen alternatieve vormen van gebruik dan in het geval van oppervlaktewater. Voor de totale waterbehoefte in koffieproductie maakt het nauwelijks verschil of het droge of natte productieproces wordt toegepast, omdat het water in het natte productieproces een zeer kleine fractie (0,34%) is van het water dat wordt gebruikt voor de plantgroei. Het effect van deze vrij kleine hoeveelheid water is echter vaak significant. Ten eerste is het blauw water (onttrokken van oppervlakte- en grondwater), dat soms schaars is. Ten tweede, het afvalwater dat in het natte productieproces wordt geproduceerd is vaak zwaar verontreinigd.

Op vergelijkbare wijze als bij koffie en thee, is de consumptie van een katoenproduct verbonden met een reeks van effecten op de waterbronnen in de landen waar het katoen wordt verbouwd en verwerkt. De studie heeft de 'watervoetafdruk' van de mondiale katoenconsumptie geschat, waarbij zowel de plaats als het karakter van de effecten zijn geïdentificeerd. Het onderzoek maakt onderscheid tussen drie soorten effect: verdamping van geïnfiltreerd regenwater voor de groei van de katoenplant (groen watergebruik), onttrekking van grond - of oppervlaktewater voor irrigatie of katoenverwerking (blauw watergebruik) en waterverontreiniging tijdens de groei of de verwerking. Het laatstgenoemde effect is gekwantificeerd in termen van het verdunningsvolume dat noodzakelijk is om de verontreiniging te assimileren. Voor de periode 1997-2001 toont het onderzoek aan dat de mondiale consumptie van katoenproducten 256 Gm^3 water per jaar vereist, waarvan ongeveer 42% blauw water is, 39% groen water en 19% verdunningswater. De effecten zijn typisch grensoverschrijdend. Ongeveer 84% van de watervoetafdruk van katoenconsumptie in de EU25-regio is buiten Europa gelokaliseerd, met belangrijke effecten in het bijzonder in India en Oezbekistan. Gezien het algemene gebrek aan afdoende waterprijsmechanismen of andere manieren om productie-informatie over te brengen, hebben katoenconsumenten weinig aansporing om de verantwoordelijkheid voor de effecten op ver afgelegen watersystemen te nemen.

Het onderzoek toont aan dat internationale handel op indirecte wijze de watergebruikefficiëntie op mondiaal niveau heeft verbeterd en heeft bijgedragen aan de oplossing van de nationale waterschaarste in sommige waterarme landen door aldaar nationale waterbronnen te besparen. Dit ging echter gepaard met verhoogde waterafhankelijkheid tussen landen. De bestaande indicatoren van watergebruik volstaan niet in het weergeven van het effect van consumptie op waterbronnen. Het wordt voorgesteld om het concept van de watervoetafdruk te hanteren om het echte watergebruik van een land te begrijpen en inzicht te krijgen in de effecten van lokale consumptie op de mondiale waterbronnen. De toekomstige onderhandelingen over verdere liberalisering van de wereldhandel zouden moeten uitgaan van het besef dat handel niet alleen een hulpmiddel van economische ontwikkeling is; het kan ook bedoeld of onbedoeld een middel zijn om de watervoetafdruk te externaliseren, waarbij de milieulasten worden verplaatst naar veraf gelegen plaatsen.

References

Albersen, P. J., Houba, H. E. D. and Keyzer, M. A. (2003) Pricing a raindrop in a process-based model: General methodology and a case study of the Upper-Zambezi. *Physics and Chemistry of the Earth* **28**: 183-192.

Allan, J. A. (1993) Fortunately there are substitutes for water otherwise our hydro-political futures would be impossible. In: *Priorities for water resources allocation and management*, pp. 13-26. ODA, London. .

Allan, J. A. (1994) Overall perspectives on countries and regions. In: *Water in the Arab World: perspectives and prognoses*, eds. P. Rogers and P. Lydon, pp. 65-100. Harvard University Press, Cambridge, Massachusetts.

Allan, J. A. (1997) Virtual water: A long term solution for water short Middle Eastern economies? In: *British Association Festival of Science*, University of Leeds: Water Issues Group, School of Oriental & African Studies, University of London.

Allan, J. A. (1998a) Virtual water: A strategic resource, global solutions to regional deficits. *Groundwater* **36**(4): 545-546.

Allan, J. A. (1998b) Watersheds and problemsheds: Explaining the absence of armed conflict over water in the Middle East. *MERIA - Middle East Review of International Affairs* **2**(1): 1-3.

Allan, J. A. (1999a) Global systems ameliorate local draughts: water food and trade. In: *Occasional Paper No 10*, London: Water Issues Study Group, SOAS, University of London.

Allan, J. A. (1999b) Water Stress and Global Mitigation: Water, Food and Trade. *Arid Land Newsletter* **45**.

Allan, J. A. (2001a) *The Middle East water question: Hydropolitics and the global economy*. London: I.B. Tauris.

Allan, J. A. (2001b) Virtual Water - economically invisible and politically silent - a way to solve strategic water problems. *International Water and Irrigation* **21**(4): 39-41.

Allan, J. A. (2002) Water resources in semi-arid regions: Real deficits and economically invisible and politically silent solutions. In: *Hydropolotics in the developing world: A Southern African perspective*, eds. A. R. Turton and R. Henwood, pp. 23-36.

Allan, J. A. and Mallat, C. (2002) Water in the Middle East: Legal, Political and Commercial Implications. The Centre of Islamic and Middle Eastern Law (CIMEL), SOAS, University of London.

Allen, R. G., Pereira, L. S., Raes, D. and Smith, M. (1998) Crop evapotranspiration - Guidelines for computing crop water requirements. *FAO Irrigation and Drainage Paper 56*, Rome, Italy: FAO.

Berkoff, J. (2003) China: the South-North Water Transfer Project - is it justified ? *Water Policy* **5**: 1-28.

Bressani, R. (2003) Coffea arabica: Coffee, coffee pulp. In: *AFRIS-2003*, Rome: Animal Feed Resources Information System, FAO.

Brown, L. R. and Halweil, B. (1998) China's water shortage could shake world food security. *World Watch Magazine*, **5**: 10-21.

Cavanagh, J. and Mander, J. (2002) *Alternatives to economic globalisation: a better world is possible*. San Francisco: The International Forum on Globalization, Berett-Koehler Publisher, Inc.

CCI (2005) Regions of US productions. *http://www.cottonusa.org/economicdata/index.cfm?ItemNumber=856*.

Chapagain, A. K. (2000) Exploring methods to assess the value of water: A case study on Zambezi basin. *Value of Water Research Report Series No. 1*, Delft, the Netherlands: UNESCO-IHE.

Chapagain, A. K. and Hoekstra, A. Y. (2003a) Virtual water flows between nations in relation to trade in livestock and livestock products. *Value of Water Research Report Series No. 13*, Delft, the Netherlands: UNESCO-IHE.

Chapagain, A. K. and Hoekstra, A. Y. (2003b) Virtual water trade: A quantification of virtual water flows between nations in relation to international trade of livestock and livestock products. *Virtual water trade: Proceedings of the International Expert Meeting on Virtual Water Trade, Value of Water Research Report Series No. 12*, ed. A. Y. Hoekstra, Delft, the Netherlands: UNESCO-IHE.

Chapagain, A. K. and Hoekstra, A. Y. (2003c) The water needed to have the Dutch drink coffee. *Value of Water Research Report Series No. 14*, Delft, the Netherlands: UNESCO-IHE.

Chapagain, A. K. and Hoekstra, A. Y. (2003d) The water needed to have the Dutch drink tea. *Value of Water Research Report Series No 15*, Delft, the Netherlands: UNESCO-IHE.

Chapagain, A. K. and Hockstra, A. Y. (2004) Water footprints of nations. *Value of Water Research Report Series No. 16*, Delft, the Netherlands: UNESCO-IHE.

Chapagain, A. K. and Hoekstra, A. Y. (submitted-a) The global component of freshwater demand and supply. *Water International*.

Chapagain, A. K. and Hoekstra, A. Y. (submitted-b) The water footprint of coffee and tea consumption in the Netherlands. *Ecologial Economics*.

Chapagain, A. K., Hoekstra, A. Y. and Savenije, H. H. G. (2005a) Saving water through global trade. *Value of Water Research Report Series No. 17*, Delft, the Netherlands: UNESCO-IHE.

Chapagain, A. K., Hoekstra, A. Y. and Savenije, H. H. G. (2005b) Water saving through international trade of agricultural products. *Hydrology and Earth System Sciences Discussions* 2: 2219-2251.

Chapagain, A. K., Hoekstra, A. Y., Savenije, H. H. G. and Gautam, R. (2005c) The water footprint of cotton consumption. *Value of Water Research Report Series No. 18*, Delft, the Netherlands: UNESCO-IHE.

Chapagain, A. K., Hoekstra, A. Y., Savenije, H. H. G. and Gautam, R. (2005d) The water footprint of cotton consumption: an assessment of the impact of worldwide consumption of cotton products on the water resources in the cotton producing countries. *Ecologial Economics* (Accepted).

Cornelis, G., van Kooten and Erwin, H. B. (2000) The ecological footprint: useful science or politics? *Ecological Economics* 32: 385-389.

Cosgrove, B. and Rijsberman, F. (2000) *World Water Vision: Making Water Everybody's Business*. London: World Water Council, Earthscan Publications Ltd.

Cotton Australia (2005) How to grow a pair of Jeans. *http://www.cottonaustralia.com.au/AC_howtogrow.htm*.

CRC (2004) NUTRIpak: A practical guide to cotton nutrition. *http://cotton.pi.csiro.au/Publicat/Agro/Nutrient/NUTRIpak.htm*.

Daveri, F., Manasse, P. and Serra, D. (2003) The Twin Effects of Globalization. *Working Paper No. 3154*, Washington DC: The World Bank.

de Fraiture, C., Cai, X., Amarasinghe, U., M., R. and Molden, D. (2004) Does International Cereal Trade Save Water? The Impact of Virtual Water Trade on Global Water Use. *Comprehensive Assessment Research Report No 4*, Sri Lanka: IWMI.

de Man, R. (2001) The global cotton and textile chain: Substance flows, actors and co-operation for sustainability, A study in the framework of WWF's Freshwater and Cotton Programme. Leiden, the Netherlands: Reinier de Man Publications.

Dicken, P. (1992) *Global shift: the internationalization of economic activity*. London: Paul Chapman Publishing Ltd.

Dubois, P. (2001) International cooperation for the development of a sustainable coffee economy. *The first Asian regional round-table on sustainable, organic and speciality coffee production, processing and marketing*, eds. K. Chapman and S. Subhadrabandhu, Chiang Mai, Thailand.

Duke, J. A. (1983) Handbook of Energy Crops, unpublished. In: *Camellia sinensis (L.) Kuntze*.

Earle, A. (2001) The Role of Virtual Water in Food Security in Southern Africa. In: *Occasional Paper No 33,* Water Issues Study Group, School of Oriental and African Studies (SOAS), University of London.

Ellwood, W. (2001) *The no-nonsense guide to globalization.* UK: New Internationalist Publications Ltd. in association with Verso.

EPA (2005) List of drinking water contaminants: Ground water and drinking water. *http://www.epa.gov/safewater/mcl.html#1.*

Falkenmark, M. (1989) The massive water scarcity now threatening Africa: why isn't it being addressed? *Ambio* **18**(2): 112-118.

Falkenmark, M. (1995) Land-water linkages: a synopsis. In: *Land and Water Integration and River Basin Management,* pp. 15-16. Rome: FAO.

Falkenmark, M. and Lundqvist, J. (1997) World freshwater problems - call for a new realism. *Background report to the 'Comprehensive assessment of the freshwater resources of the world',* Stockholm, Sweden: Stockholm Environment Institute.

Falkenmark, M., Lundqvist, J. and Widstrand, C. (1989) Macro-scale water scarcity requires micro-scale approaches: aspects of vulnerability in semi-arid development. *Natural Resources Forum*: 258-267.

Falkenmark, M. and Rockström, J. (2004) *Balancing water for humans and nature: the new approach in ecohydrology.* London: Earthscan.

FAO (1999) Irrigation in Asia in Figures. *Water Reports 18*, Rome, Italy: Food and Agriculture Organization of the United Nations.

FAO (2003a) AQUASTAT 2003. *ftp://ftp.fao.org/agl/aglw/aquastat/aquastat2003.xls.*

FAO (2003b) CLIMWAT database. *http://www.fao.org/ag/AGL/aglw/climwat.stm.*

FAO (2003c) CROPWAT model. *http://www.fao.org/ag/AGL/aglw/cropwat.htm.*

FAO (2003d) FAO Statistical Databases. *http://faostat.fao.org/.*

FAO (2003e) FAOCLIM: a CD-ROM with world-wide agroclimatic data. *http://www.fao.org/sd/2001/EN1102_en.htm.*

FAO (2003f) Technical conversion factors for agricultural commodities. Rome, Italy: Food and Agriculture Organization of the United Nations.

FAO (2005) Review of global agricultural water use per country, crop water requirements. *http://www.fao.org/ag/agl/aglw/aquastat/water_use/index4.stm.*

FAOSTAT (2004) FAO Statistical Databases: Food Balance Sheets. *http://faostat.fao.org/.*

FAOSTAT data (2005) FAO Statistical Databases, last updated February 2005. *http://faostat.fao.org/default.jsp.*

Gallopín, G. C. and Rijsberman, F. (2000) Three global scenarios. *International Journal of Water* **1**(1): 16-40.

Gillham, F. E. M., Bell, T. M., Arin, T., Matthews, T. A., Rumeur, C. L. and A.B., H. (1995) Cotton production prospects for the next decade. *World Bank Technical Paper Number 287*, Washington DC: the World Bank.

Glantz, M. H. (1998) Creeping environmental problems in the Aral Sea basin. In: *Central Eurasian water crisis: Caspian, Aral and dead seas*, eds. I. Kobori and M. H. Glantz, New York: United Nations University Press.

Gleick, P. H. (1991) Water and conflict. *International Security* **18**: 79-112.

Gleick, P. H., ed. (1993) *Water in crisis: A guide to the world's fresh water resources.* Oxford, UK: Oxford University Press.

Gleick, P. H., Wolff, G., Chalecki, E. L. and Reyes, R. (2002a) Globalization and international trade of water. In: *The world's water: the biennial report on freshwater resources 2002-2003*, eds. P. H. Gleick, W. C. G. Burns, E. L. Chalecki, M. Cohen, K. K. Cushing, A. S. Mann, R. Reyes, G. H. Wolff and A. K. Wong, pp. 33-56. Washington: Island Press.

Gleick, P. H., Wolff, G., Chalecki, E. L. and Reyes, R. (2002b) *The new economy of water: the risks and benefits of globalization and privatization of fresh water.* California: Pacific Institute for Studies in Development, Environment, and Security.

Greenaway, F., Hassan, R. and Reed, G. V. (1994) An empirical analysis of comparative advantage in Egyptian agriculture. *Appl. Eco.* **26**: 649-657.

GTZ (2002a) Post harvesting processing: Coffee drying.
  http://www.venden.de/postharvestprocessing.htm.
GTZ (2002b) Post harvesting processing: Coffee waste water.
  http://www.venden.de/postharvestprocessing.htm.
GTZ (2002c) Post harvesting processing: Facts and figures.
  http://www.venden.de/postharvestprocessing.htm.
Haddadin, M. J. (2003) Exogenous water: A conduit to globalization of water resources. *Virtual water trade: Proceedings of the International Expert Meeting on Virtual Water Trade, Value of Water Research Report Series No. 12*, ed. A. Y. Hoekstra, Delft, the Netherlands: UNESCO-IHE.
Hall, M., Dixon, J., Gulliver, A. and Gibbon, D., eds. (2001) *Farming Systems and Poverty: Improving farmer's livelihoods in a changing world.* Rome and Washington: FAO and World Bank
Herendeen, R. A. (2000) Ecological footprint is a vivid indicator of indirect effects. *Ecological Economics* **32**: 357-358.
Herendeen, R. A. (2004) Energy analysis and EMERGY analysis - A comparison. *Ecol. Model.* **178**: 227-237
Hicks, P. A. (2001) Postharvest processing and quality assurance for speciality/organic coffee products. *The first Asian regional round-table on sustainable, organic and speciality coffee production, processing and marketing*, eds. K. Chapman and S. Subhadrabandhu, Chiang Mai, Thailand.
Hoekstra, A. Y. (1998) *Perspectives on water: an integrated model-based exploration of the future.* Utrecht, the Netherlands: International Books.
Hoekstra, A. Y. (2003) Virtual water: An introduction. *Virtual water trade: Proceedings of the International Expert Meeting on Virtual Water Trade, Value of Water Research Report Series No 12*, ed. A. Y. Hoekstra, Delft, The Netherlands: UNESCO-IHE.
Hoekstra, A. Y. and Chapagain, A. K. (2004a) Eén kopje koffie kost gemiddeld 140 liter water. *H2O* **37**(5): 36-37.
Hoekstra, A. Y. and Chapagain, A. K. (2004b) Water Footprint - A Consumption-based indicator of water pressure. *Background Paper to Chapter 4 of "Let it reign: The new water paradigm for global food security, a report to the UN Commission on Sustainable Development 13*: Swedish International Development Corporation Agency.
Hoekstra, A. Y. and Chapagain, A. K. (2005a) De water-voetafdruk van de Nederlanders en de wereldbevolking. *H2O* **38**(4): 37-41.
Hoekstra, A. Y. and Chapagain, A. K. (2005b) The effect of international trade in agricultural products on national water demand and scarcity, with examples for Morocco and the Netherlands. *Conference on water on the occasion of 400 years of international relations between Netherlands and Morocco*, Marrakech, Morocco.
Hoekstra, A. Y. and Chapagain, A. K. (2005c) Water footprints of nations: water use by people as a function of their consumption pattern. *Water Resources Management* (Accepted for publication).
Hoekstra, A. Y., Hoekstra, A. Y., Savenije, H. H. G. and Chapagain, A. K. (2002) Water value flows: A case study in Zambezi basin. *Value of Water Research Report Series No. 2*, Delft, the Netherlands: UNESCO-IHE.
Hoekstra, A. Y. and Hung, P. Q. (2002) Virtual water trade: A quantification of virtual water flows between nations in relation to international crop trade. *Value of Water Research Report Series No. 11*, Delft, the Netherlands: UNESCO-IHE.
Hoekstra, A. Y. and Hung, P. Q. (2005) Globalisation of water resources: International virtual water flows in relation to crop trade. *Global Environmental Change* **15**(1): 45-56.
Hoekstra, A. Y., Savenije, H. H. G. and Chapagain, A. K. (2001) An integrated approach towards assessing the value of water: A case study on the Zambezi basin. *Integrated Assessment* **2**(4): 199-208.
Hoekstra, A. Y., Savenije, H. H. G. and Chapagain, A. K. (2003) The value of rainfall: up-scaling economic benefits to the catchment scale. In: *SIWI Seminar 'Towards catchment hydrosolidarity in a world of uncertainties*, pp. 63-68. Stockholm.

ICO (2003) Website of International Coffee Organization. *http://www.ico.org.*

IFA, IFDC, IPI, PPI and FAO (2002) Fertilizer use by crop. *http://www.fertilizer.org/ifa/statistics/crops/fubc5ed.pdf.*

ITC (1999) PC-TAS version 1995-1999 in HS or SITC, CD-ROM. Geneva: International Trade Centre.

ITC (2004) PC-TAS version 1997-2001 in HS or SITC, CD-ROM. Geneva: International Trade Centre.

Jeroen, C. J. M., van deli, B. and Verbruggen, H. (1999) Spatial sustainability, trade and indicators: an evaluation of the 'ecological footprint'. *Ecological Economics* **39**: 61-72.

Kulshreshtha, S. N. (1993) World water resources and regional vulnerability: impact of future changes. In: *Research Report RR-93-10*, Laxenburg, Austria: International Institute for Applied Systems Analysis

L'Amyx (2003) Tea history and legend. *http://www.lamyx.com/articles/tealegend.html.*

Lenzen, M. and Murray, S. A. (2001) A modified ecological footprint method and its application to Australia. *Ecological Economics* **37**: 229-255.

López-Ortiz, A. and Owen, D. (2003) Frequently asked questions about coffee. In: *The Coffee and Caffeine FAQ. http://coffeefaq.com/coffaq1.htm*

Ma, J., Hoekstra, A. Y., Wang, H., Chapagain, A. K. and Wang, D. (2005) Virtual versus real water transfer within China. *Philosophical Transactions: Biological Sciences* DOI: 10.1098/rstb.2005.1644 (FirstCite Early Online Publishing).

Nakayama, M. (2003) Implications of virtual water concept on management of international water systems – cases of two Asian international river basins. *Virtual water trade: Proceedings of the International Expert Meeting on Virtual Water Trade, Value of Water Research Report Series No. 12*, ed. A. Y. Hoekstra, Delft, The Netherlands: UNESCO-IHE.

Oki, T. and Kanae, S. (2004) Virtual water trade and world water resources. *Water Science & Technology* **49**(7): 203–209.

Oki, T., Sato, M., Kawamura, A., Miyake, M., Kanae, S. and Musiake, K. (2003) Virtual water trade to Japan and in the world. *Virtual water trade: Proceedings of the International Expert Meeting on Virtual Water Trade, Value of Water Research Report Series No. 12*, ed. A. Y. Hoekstra, Delft, the Netherlands: UNESCO-IHE.

Opschoor, H. (2000) The ecological footprint: measuring rod or metaphor? *Ecological Economics* **32**: 363-365.

Oxfam (2003) *EU Hypocrisy Unmasked: Why EU Trade Policy Hurts Development.* Brussels: Oxfam International EU Advocacy Office.

Pereira, L. S., Cordery, I. and Iacovides, I. (2002) Coping with water scarcity. In: Paris: International Hydrological Programme, UNESCO.

Postel, S. (1992) *Last oasis: Facing water scarcity.* New York: W.W Norton & Company.

Postel, S. L., Daily, G. C. and Ehrlich, P. R. (1996) Human appropriation of renewable fresh water. *Science* **271**: 785-788.

Proto, M., Supino, S. and Malandrino, O. (2000) Cotton: a flow cycle to exploit. *Industrial Crops and Products* **11**(2-3): 173-178.

Qian, Z. Y., Zhang, W. Z., Lin, B. N. and Sun, X. T. (2002) Comprehensive report of strategy on water resources for China's sustainable development. pp. 38-41. Beijing, China: China Water and Hydropower Press.

Raskin, P., Hansen, E. and Margolis, R. (1995) Water and sustainability: a global outlook. *Polestar Report Series No 4*, Boston, USA: Stockholm Environment Institute.

Rees, W. E. (1992) Ecological footprints and appropriated carrying capacity: what urban economics leaves out. *Environ. Urban.* **4**(2): 121-130.

Rees, W. E. (1996) Revisiting carrying capacity: area based indicators of sustainability. *A Journal of Interdisciplinary Studies* **17**(3).

Ren, X. (2000) Development of environmental performance indicators for textile process and product. *Journal of Cleaner Production* **8**(6): 473-481.

Renault, D. (2003) Value of virtual water in food: Principles and virtues. *Virtual water trade: Proceedings of the International Expert Meeting on Virtual Water Trade, Value of*

Water Research Report Series No 12, ed. A. Y. Hoekstra, Delft, the Netherlands: UNESCO-IHE.

Rennen, W. and Martens, P. (2003) The globalisation timeline. *Integrated Assessment* **4**(3): 137-144.

Roast and Post (2003) From tree to cup, processing. *http://www.realcoffee.co.uk/Article.asp?Cat=TreeToCup&Page=3.*

Rockström, J. and Gordon, L. (2001) Assessment of green water flows to sustain major biomes of the world: implications for future ecohydrological landscape management. *Phys. Chem. Earth* **B**(26): 843-851.

Rosegrant, M. W. and Ringler, C. (1999) Impact on food security and rural development of reallocating water from agriculture. Washington DC: IFPRI.

Rosenblatt, L., Meyer, J. and Beckmann, E. (2003) *Koffie: Geschiedenis, teelt, veredeling, met 60 heerlijke koffierecepten.* Abcoude, the Netherlands: Fontaine Uitgevers.

Savenije, H. H. G. (2000) Water scarcity indicators; the deception of numbers. *Phys. Chem. Earth (B)* **25**(3): 199-204.

Savenije, H. H. G. (2004) The role of green water in food production in sub-saharan Africa. *http://www.wca-infonet.org/cds_upload/documents/1352.Role_of_green_water.pdf.*

Savenije, H. H. G. and van der Zaag, P. (2002) Water as an economic good and demand management: paradigms with pitfalls. *Water International* **27**(1): 98-104.

Sciona (2003) Storm in a tea cup. *http://www.sciona.com/coresite/nutrition/articles/storm_teacup.htm.*

Senbel, M., McDaniels, T. and Dowlatabadi, H. (2003) The ecological footprint: a non-monetary metric of human consumption applied to North America. *Global Environmental Change* **13**: 83-100.

Seuring, S. (2004) Integrated chain management and supply chain management: Comparative analysis and illustrative cases. *Journal of Cleaner Production* **12**: 1059-1071.

Shiklomanov, I. A. (2000) Appraisal and assessment of world water resources. *Water International* **25**(1): 11-32.

Shuval, H. (1998) A revaluation of conventional wisdom on water security, food security, and water stress in arid countries in the Middle East. *Water workshop: Averting a water crisis in the Middle East - make water a medium of cooperation rather than conflict, Mar 18, 1998*, Geneva: Green Cross and UNESCO International Hydrological Program.

Silvertooth, J. C., Navarro, J. C., Norton, E. R. and Galadima, A. (2001) Soil and plant recovery of labeled fertilizer nitrogen in irrigated cotton. In: *Arizona Cotton Report*: University of Arizona.

SIWI, IFPRI, IUCN and IWMI (2005) Let it reign: the new water paradigm for global food security. *Working report to CSD-13*, Stockholm: Stockholm International Water Institute.

Soth, J., Grasser, C. and Salerno, R. (1999) The impact of cotton on fresh water resources and ecosystems: A preliminary analysis. Gland, Switzerland: WWF.

Sovrana (2003) Website of Sovrana Trading Corporation. *http://sovrana.com/roasting.htm.*

Suo, L. S. (2004) River management and ecosystem conservation in China. In: *Ninth international symposium on river sedimentation*, pp. 3-10. Yichang, China: Ministry of Water Resources.

The Fragrant Leaf (2003) Basic tea brewing and storage. *http://www.thefragrantleaf.com/teaprepbasbr.html.*

Turton, A. R. (1999) Precipitation, people, pipelines and power: Towards a "virtual water" based political ecology discourse. *MEWREW Occasional Paper, OCC11*, Water Issues Study Group, SOAS, University of London.

Turton, A. R. (2002) A strategic decision-makers guide to virtual water. African Water Issues Research Unit (AWIRU), Centre for International Political Studies (CIPS), Pretoria University.

Twinings (2003a) Black tea manufacture. *http://www.twinings.com/en_int/tea_production/orthodox.html*

Twinings (2003b) Tea reaches Europe.
  http://www.twinings.com/en_int/history_tradition/europe.asp.
UNCTAD (2005a) Cotton uses. *http://r0.unctad.org/infocomm/anglais/cotton/uses.htm.*
UNCTAD (2005b) Planting and harvesting times for cotton, by producing country.
  http://r0.unctad.org/infocomm/anglais/cotton/crop.htm.
UNEP (2002) Global environment outlook 3: past, present and future perspectives. In:
  London, UK: Earthscan Publications Ltd.
UNEP IE (1996) Cleaner production in textile wet processing: a workbook for trainers. In:
  Paris: United Nations Environment Programme: Industry and Environment.
UNESCO-WWAP (2003) *Water for people, water for life - United Nations World Water
  Development Report.* Paris: UNESCO Publishing.
USDA (2004) Cotton: World markets and trade.
  http://www.fas.usda.gov/cotton/circular/2004/07/CottonWMT.pdf.
USDA/NOAA (2005a) Cotton - World supply and demand summary.
  http://www.tradefutures.cc/education/cotton/worldsd.htm.
USDA/NOAA (2005b) Major world crop areas and climatic profiles.
  http://www.usda.gov/agency/oce/waob/mississippi/MajorWorldCropAreas.pdf.
USEPA (1996) Best management practices for pollution prevention in the textile industry-
  1996. *http://www.e-textile.org.*
Van den Bergh, J. C. J. M. and Verbruggen, H. (1999) Spatial sustainability, trade and
  indicators: An evaluation of the 'ecological footprint. *Ecologial Economics* **29**: 61-
  72.
Van Kooten, G. C. and Bulte, E. H. (2000) The ecological footprint: useful science or politics.
  Ecologial Economics **32**: 385-389.
Van Wieringen, S. (2001) Alles over koffie. Culemborg, the Netherlands.: Fair Trade.
Visvanathan, C., Kumar, S. and Han, S. (2000) Cleaner production in textile sector: Asian
  scenario. In: *National Workshop on Sustainable Industrial Development through
  Cleaner Production*, Colombo, Sri Lanka.
Vörösmarty, C. J., Green, P., Salisbury, J. and Lammers, R. B. (2000) Global water resources:
  Vulnerability from climate change and population growth. *Science* **289**: 284-288.
Wackernagel, M. and Jonathan, L. (2001) *Measuring sustainable development: Ecological
  footprints.* Mexico: Centre for Sustainability Studies, Universidad Anahuac de
  Xalapa.
Wackernagel, M., Onisto, L., Bello, P., Linares, A. C., Falfan, I. S. L., Garcia, J. M.,
  Guerrero, A. I. S. and Guerrero, M. G. S. (1999) National natural capital accounting
  with the ecological footprint concept. *Ecologial Economics* **29**: 375-390.
Wackernagel, M., Onisto, L., Linares, A. C., Falfan, I. S. L., Garcia, J. M., Guerrero, I. S. and
  Guerrero, M. G. S. (1997) *Ecological footprints of nations: How much nature do
  they use? How much nature do they have?* Mexico: Centre for Sustainability
  Studies, Universidad Anahuac de Xalapa.
Wackernagel, M. and Rees, W. (1996) *Our ecological footprint: Reducing human impact on
  the earth.* Gabriola Island, B.C., Canada: New Society Publishers.
Wang, M. R., Li, C. H., Zhang, W. and Ren, L. L. (2001) Impacts of human activities on river
  runoff in North China. *Journal of Hohai University Natural Science* **29**: 13-18.
Warner, J. (2003) Virtual water – virtual benefits? Scarcity, distribution, security and conflict
  reconsidered. *Virtual water trade: Proceedings of the International Expert Meeting
  on Virtual Water Trade, Value of Water Research Report Series No 12*, ed. A. Y.
  Hoekstra, Delft, the Netherlands: UNESCO-IHE.
Watkins, K. and Fowler, P. (2002) Rigged rules and double standards: Trade globalisation
  and fight against poverty. Oxfam & Make Trade Fair.
WB (1999) Pollution prevention and abatement handbook 1998: Toward cleaner production.
  Washington DC: World Bank.
WHO/UNICEF (2000) Global Water Supply and Sanitation Assessment 2000 Report.
  Geneva: World Health Organization/United Nations Children's Fund.

Wichelns, D. (2001) The role of 'virtual water' in efforts to achieve food security and other national goals, with an example from Egypt. *Agricultural Water Management* **49**(2): 131-151.

Wichelns, D. (2003) The role of public policies in motivating virtual water trade, with an example from Egypt. *Virtual water trade: Proceedings of the International Expert Meeting on Virtual Water Trade, Value of Water Research Report Series No 12*, ed. A. Y. Hoekstra, Delft, the Netherlands: UNESCO-IHE.

Wichelns, D. (2004) The policy relevance of virtual water can be enhanced by considering comparative advantages. *Agricultural Water Management* **66**(1): 49-63.

WTO (2004) Statistical database. World Trade Organization.

WWF (1999) The impact of cotton on fresh water resources and ecosystem: a preliminary synthesis. *http://www.panda.org/downloads/freshwater/impact_long.doc.*

WWF (2003) Thirsty Crops. Our food and clothes: Eating up nature and wearing out the environment?, Living waters: conserving the source of life, the Netherlands: WWF.

WWF (2004) *Living water planet report 2004*. WWF.

Yang, H., Reichert, P., Abbaspour, K. C. and Zehnder, A. J. B. (2003) A water resources threshold and its implications for food security. *Virtual water trade: Proceedings of the International Expert Meeting on Virtual Water Trade, Value of Water Research Report Series No 12*, ed. A. Y. Hoekstra, Delft, the Netherlands: UNESCO-IHE.

Yang, H. and Zehnder, A. J. B. (2002) Water scarcity and food import: A case study for Southern Mediterranean countries. *World Development* **30**(8): 1413–1430

Zimmer, D. and Renault, D. (2003) Virtual water in food production and global trade: Review of methodological issues and preliminary results. *Virtual water trade: Proceedings of the International Expert Meeting on Virtual Water Trade, Value of Water Research Report Series No. 12*, ed. A. Y. Hoekstra, Delft, the Netherlands: UNESCO-IHE.

Acknowledgements

With the background of civil engineering, to embark on a research in the field of interdisciplinary science was like dreaming to climb the Mount Everest. With dream came the helpful hands who have generously contributed their time and energy in one way or another to meet the peak. Though it is not possible to mention each one here, I would like to express my sincere thanks to all of them.

Though none of you were less important, some of you were exceptionally driven by my desire to keep on climbing. When I read the story Animal Farm by George Orwell, it gave me a life time impression how inequality exists among equals. Worth mentioning is my supervisor for his role as a guide, friend and promotor Professor Arjen Hoekstra. I still remember when I was looking for an interesting research topic for my MSc then I first met him in 1999 at IHE and that was the turning point in my life. I chose to explore methods to assess the value of water along with the hydrological cycle and that ultimately opened a new dimension in me and the current research is the culmination of that interest. I admire his approach in guiding a research to the extent that I followed him not only at Delft where he was at walking distance of 10 minutes from my hostel but also to Twente which is at 3 hours travel time by Train, and even further to Zimbabwe with Flight time of more than a day. My sincere gratitude goes to him.

Combining economics with hydrology was a bit tough task for me. I got constructive and diverse comments from our friends in economics. This has definitely contributed to improving the quality of the thesis. Though at times my ignorance and little knowledge of economics used to create a mess, my other promotor Professor Huub Savenije was always there to show me the right way and the basics of economics when it comes to water resources management. My sincere gratitude goes to Huub for his guidance, patience and faith in my abilities without which it would have been impossible to complete this research.

Throughout the Journey I have had pleasure of getting in touch with some unique set of peoples with different opinions, ideas, and perspectives. Every touch left unique impacts in my life. I may not recognise them explicitly one by one, but as a whole it has transformed my personality to what I'm today. I recall Frank Jaspers, Pieter van der Zaag, and the Water-Net community for Harare experiences. Special thank goes to Annemiek, Vandana, Saroj, Patrick and Eline for their moral support while I was at the Delft.

This research has highly benefited from the individual efforts of several MSc students namely Zhang Dunquiang, Jing Ma, Xiuying Dong, Abbas Iglal, Thewodros Gebre, Mesfin Mekonnen and Rajani Gautam. They deserve my heartfelt thanks not only for being part of the research but also for the invaluable discussions that we had during our weekly meetings. In the course of their research I even had opportunities to discuss the fundamentals of my research with their examiners and I found most of their suggestions beneficial to my thesis as a whole. The constructive comment from Professor Huub Gijzen has influenced this research to an extent that I have included a whole new dimension of the effect of pollution on the global water resources from consumption of goods. People say that a picture is worth thousand words, however it should be conveying its message without the expense of thousands words. I am thankful to Daniel Schotanus for providing skill necessary to produce maps using Geographical Information System.

I received extraordinary supports from the colleagues of the Department of Environment Resources, the Department of Water Resources Management, and the PhD research community at IHE. I was fortunate to see 8 batches of Master of Engineering students from all over the world at IHE. I shared a lot with this unique community, not only my subject but also my personal life. When I look back, they were the one who never let me down and supported in all respect. Thanks for the friendship which I would cherish as a part of my life even beyond this accomplishments.

Getting a financial support for a research in one's own field of interest is the biggest obstacles that most of the promising researchers from developing countries face. I am thankful to the Dutch National Institute of Public Health and the Environment (RIVM) for financing the research. I thank NOVIB for contributing to the coffee and tea case study.

In February 2005, a tsunami like event passed through me, I lost my father. My family went back to Nepal and I was left blank. At one point in time I almost decided to leave the time-bound research. I got in contact with my mother and she encouraged me to complete my research whatever may come. In my childhood, I learnt how to write the alphabets from her at home. Now these letters seem insufficient to formulate meaningful words to show my feelings to her. In these darkest hours of my life, I got strongest support from one of the dearest and nearest to me, my wife Indira. She shared those moments with hope and enthusiasm with me. That gave me the extra push sufficient enough to get ready for the final climb. While writing this acknowledgement it is already 3 years and 3 months that we're living in a family fragmented physically but united emotionally. I have learnt no words so far to express what I owe to my beloved wife Indira.

Finally, at the top of everything, there are two most beautiful angels, my sons Akshu and Ansu. They spent a large part of their childhood without their father when they needed most. If there is a cost for everything, then the cost of this journey is what my innocent ones paid.

About the author

Ashok Chapagain was born on 28 January 1965 in a remote hilly village Dingla in Bhojpur district in Nepal. He completed his basic education in Biratnagar, the second largest city in Nepal. He joined the Indian Institute of Technology Roorkee (then University of Roorkee, India) in 1985 to pursue the course of Bachelor of Civil Engineering and received the degree with honours in 1989. After a career as an irrigation engineer in Nepal for 9 years, he joined the Master of Science programme in 1998 at UNESCO-IHE (then IHE), the Netherlands. He received his Master of Science degree with distinction in Water and Environmental Resources Management in February 2000. His Master of Science thesis is about exploring methods to assess the value of water with an integrated approach which is then applied to the Zambezi river basin. Since September 2002 he is at UNESCO-IHE to pursue a PhD research programme carried out under full time construction.

He has been working in the field of irrigation and river training works in Nepal as a government engineer at various capacities since 1989. He has successfully implemented more than 40 small scale irrigation projects (ranging from 100 to 2000 hectares) in different parts of Nepal. He has 5 years of research exposure in the field of water resources management in the Netherlands and abroad and has successfully guided a number of Master of Science students during this period. His research at UNESCO-IHE has culminated in a number of publications: 9 papers in scientific journals (Chapagain and Hoekstra, submitted-a; submitted-b; Chapagain et al., 2005b; Chapagain et al., 2005d; Hoekstra and Chapagain, 2004a; 2005a; 2005c; Hoekstra et al., 2001; Ma et al., 2005), 12 papers in the proceeding of different conferences and technical reports (Chapagain, 2000; Chapagain and Hoekstra, 2003a; 2003b; 2003c; 2003d; 2004; Chapagain et al., 2005a; Chapagain et al., 2005c; Hoekstra and Chapagain, 2004b; 2005b; Hoekstra et al., 2002; Hoekstra et al., 2003). Five of these publications (1 in journal and 4 as reports) are related to his research during his Master of Science programme. Remaining publications carry significant contribution to this dissertation. He feels that his current study dealing with issues of virtual water trade adds a global dimension to the value-flow-cycle of water in a basin which, most of the time people think, is limited to the local hydrological cycle in the concerned basin. In his opinion, there exists no ready made single recipe to solve any problem, however broadening the horizon by bringing the hidden issues to the surface surely helps to find a more suitable one sooner.

For Product Safety Concerns and Information please contact our EU
representative GPSR@taylorandfrancis.com Taylor & Francis Verlag GmbH,
Kaufingerstraße 24, 80331 München, Germany

Printed and bound by CPI Group (UK) Ltd, Croydon, CR0 4YY
01/05/2025
01858471-0001